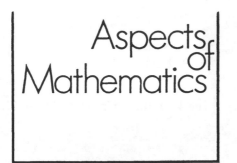

Aspects of Mathematics

Edited by Klas Diederich

*A Publication of the Max-Planck-Institut für Mathematik, Bonn

Volumes of the German-language subseries "Aspekte der Mathematik" are listed at the end of the book.

D. V. Anosov
A. A. Bolibruch

The Riemann-Hilbert Problem

A Publication from the
Steklov Institute of Mathematics

Adviser: Armen Sergeev

Professor D. V. Anosov
Professor A. A. Bolibruch
Steklov Institute of Mathematics
Vavilova 42
117966 Moscow/CIS
Russia

Die Deutsche Bibliothek – CIP-Einheitsaufnahme

Anosov, D. V.:
The Riemann Hilbert problem: a publication from
the Steklov Institute of Mathematics / D. V. Anosov;
A. A. Bolibruch.
    (Aspects of mathematics: E; Vol. 22)
    ISBN 978-3-322-92911-2   ISBN 978-3-322-92909-9 (eBook)
    DOI 10.1007/978-3-322-92909-9

NE: Bolibruch, A. A.:; Aspects of mathematics / E

Mathematics Subject Classification: 34A20

Vieweg is a subsidiary company of the Bertelsmann Publishing Group International.

Cover design: Wolfgang Nieger, Wiesbaden

Printed on acid-free paper

ISSN 0179-2156
ISBN 978-3-322-92911-2

Preface

This book is devoted to Hilbert's 21st problem (the Riemann-Hilbert problem) which belongs to the theory of linear systems of ordinary differential equations in the complex domain. The problem concerns the existence of a Fuchsian system with prescribed singularities and monodromy. Hilbert was convinced that such a system always exists. However, this turned out to be a rare case of a wrong forecast made by him. In 1989 the second author (A.B.) discovered a counterexample, thus obtaining a negative solution to Hilbert's 21st problem.[1]

After we recognized that some "data" (singularities and monodromy) can be obtained from a Fuchsian system and some others cannot, we are enforced to change our point of view. To make the terminology more precise, we shall call the following problem the Riemann-Hilbert problem for such and such data: does there exist a Fuchsian system having these singularities and monodromy? The contemporary version of the 21st Hilbert problem is to find conditions implying a positive or negative solution to the Riemann-Hilbert problem.

In this book we consider only (the contemporary version of) the classical 21st Hilbert's problem and only mention (of course, with due references) various modifications, generalizations and related problems. We mention all known results on the classical problem, both positive and negative, and prove some of them. We simply do not have enough place to prove all of them, but the samples we explain in detail include the most important cases and seem to provide a good feeling of the whole picture.

The problem under consideration is of global character, but in order to study it one needs some local theory (a theory describing the behavior of solutions near a singular point). There is a well-known local theory which goes back to Fuchs and Poincaré and can be found in such textbooks as those by Coddington-Levinson [CoLe], or Hartman [Ha](we need only a simpler part of this theory dealing with the so-called regular singularities). For our purposes this theory has to be supplemented by a new local theory due to Levelt. Our book contains the exposition of both theories inasmuch as we need them.

In 1908 Plemelj obtained a positive solution to the problem similar to Hilbert's

---

[1]In the preface we dwell on the history only inasmuch as it helps us to describe the content of the book. Introductory chapter contains more remarks on the history (which was somewhat fanciful), but complete description of the history was not our goal.

21st problem in its original form, but concerning the so-called regular systems instead of Fuchsian ones. This was a remarkable achievement, although it does not mean a solution to Hilbert's 21st problem, because the class of regular systems is broader than the class of Fuchsian systems. However, his theorem is useful even if one is interested only in Fuchsian systems – almost all positive results on Hilbert's 21st problem are obtained in the following way: one takes a regular system provided by Plemelj and tries to modify it so that it becomes Fuchsian with the same singularities and monodromy. In reality, not only the statement of Plemelj's theorem is used, but sometimes also some details from its proof – the very proof we give here. This proof is different from Plemelj's original proof, but goes back to Röhrl (1957) [Rö] and takes into account some improvements invented later. An essential "ingredient" of this proof is the use of the Birkhoff-Grothendieck theorem about the complex analytic vector bundles over the Riemann sphere. We give (with minor modifications) the elementary proof of the latter theorem developed recently by J.Leiterer [Lei]. This makes our exposition self-contained, modulo more or less standard background.

Here follows some information on this background. The reader must be acquainted with standard ("basic") courses on linear algebra (including Jordan normal form of matrices), ordinary differential equations (we need general properties of solutions to linear systems), and the theory of functions of the complex variable. Usually in the basic course of the latter more attention is paid to the "single-valued" functions than to "multivalued" ones, whereas "multivalued" functions are important for our purposes. However, we need not a "deep general theory" of such functions (whatever that means), but rather a good "feeling" of such things as branching of elementary functions, analytic continuation, "complete" analytic function, Riemann surface. Usually this is included in a basic course (although often – in a formal way) and of course can be found in many textbooks. One must know what the universal covering surface and the deck transformations are. We shall use two functions of matrices – exponential and logarithm. Although the theory of ordinary differential equations in the complex domain is rather extensive, we need only a few facts of it. It may happen that the reader's knowledge of ordinary differential equations is restricted to the real domain only. We hope that several remarks in the introductory chapter will help such a reader to adopt to a complex point of view. At the beginning our exposition is detailed, later it becomes more succinct – we hope that by that time the reader gets some practice in this field and becomes more mature.

This book includes without big changes the preprint "An introduction to Hilbert's 21st problem" [An], published by the first author (D.A.) in 1991. This preprint contained: the general introduction (which is extended here); the description of the local theory (which we reproduce here with minor changes); the first counterexample to Hilbert's 21st problem (the exposition in the preprint is somewhat improved compared to the original exposition given by the second author); proofs of the theorems of Plemelj and Birkhoff-Grothendieck. The first author wrote also the section on the Bessel equation. All other text was written by the second author. Of course, we planned and discussed it together.

The above mentioned preprint was written during the visit of the first author to the Inst. of Math. and Appl., Penn State Univ. and City Univ. of New York. In this preprint the first author already expressed his thanks to a number of persons who helped him at his work on this preprint and the entire faculty and staff of the IMA, PSU and CUNY for their hospitality. Big part of this book was written in Moscow where we both work at the Steklov Math. Inst., and the second author finished the work on this book during his visit to Max-Planck-Institut für Math. and to the University of Nice. He would like to thank the staff of MPIM and UN for hospitality and excellent working conditions. He is very grateful to V.A.Golubeva and A.V.Chernavskii , who introduced him to the theory of Fuchsian equations and the Riemann-Hilbert problem.

D.V.Anosov

A.A.Bolibruch

Contents

1 Introduction

1.1 Educational notes

When dealing with the system of ODE in a complex domain

$$\frac{dy^j}{dt} = f^j(x, y^1, \ldots, y^p), \quad j = 1, \ldots, p \tag{1.1.1}$$

one assumes that f^j are holomorphic (i.e. single-valued analytic) in some domain G and asks for analytic solution $y^1(x), \ldots, y^p(x)$. The local existence theorem (for brevity, we include here uniqueness and analytic dependence on initial data and parameters (if there are any)) looks like the corresponding theorem in a real domain. The most well-known proof of the latter is obtained by rewriting the system as a system of integral equations which is solved using iterations. A careful analysis of this proof reveals that it works in a complex domain as well. (We work in a small disk on the x-plane containing the initial value x_0 of the independent variable. Integration is performed along linear segments connecting x_0 to the "current" x. These integrals are estimated literally in the same way as in the real case). At some points it even becomes simpler. If a sequence of analytic functions converges uniformly in a domain on the x-plane, then its limit is an analytic function. In fact, one does not even need to assume a priori that the convergence is uniform; but in our case the proof of the convergence is based on the estimates which imply the uniform convergence in a sufficiently small disk. (For brevity, we say "function" instead of "a system of p functions"; one can also have in mind "a vector function"). So (1.1.1) has an analytic solution. Now, as regards to the uniqueness, it is easy to see (differentiating (1.1.1)) that two solutions having the same initial data must have the same derivatives of all orders. Being analytic, they must coincide. As regards to the dependence on initial data and parameters, we have an uniformly convergent sequence of functions holomorphic with respect to x, these data and parameters, so the limit is also holomorphic with respect to them (we again refer to the analyticity of the limit functions, only this time we deal with a function of several complex variables). Compare this easy argument with the situation in the real domain. In the latter case the un iform convergence does not imply any smoothness of the limit function. As regards to its dependence in x, smoothness follows immediately from the integral equation, so this is also easy, but as regards to the dependence on the initial data and parameters, one needs some extra considerations. (It is true that one can avoid them by using a simplest version of the implicit function theorem in Banach spaces, but it is not so popular).

Another way to prove the local existence (etc.) theorem in the complex-analytic case is provided by majorants. This is an "essentially complex-analytic " method of a broad value (its applications by no means reduce to this theorem or even to the entire theory of DE). But as regards to the theorem under consideration, it does not give more than the "integral equations plus iterations" method which may well happen to be more familiar to the reader.

Being analytic in some disk, solutions to (1.1.1) can be continued analytically. By the well-known "principle of preservation of analytic identities under the analytic continuation", the "continued" $y^1, \ldots, y^p$ remains to be a solution to (1.1.1), if $(x, y^1, \ldots, y^p)$ does not leave the domain G during the process of continuation. One must have in mind that it may happen that the solution admits the process of analytic continuation during which $(x, y^1, \ldots, y^p)$ leaves G; then the elements of function thus obtained need not be solutions, as it may happen that the right hand side of (1.1.1) does not admit an analytic continuations for such values of $(x, y^1, \ldots, y^p)$. It may happen also that after leaving G our $(x, y^1, \ldots, y^p)$ will return later to G; then we get a new element of an analytic function for which one can ask whether it will be a solution to (1.1.1). There are no reasons why it must be – it may be and it may not be. And if it will be a solution, it may be so that this solution will be different from which we started and cannot be connected to this solution via a process of analytic continuation inside G.

In the theory of real ODE there is another process of continuation – continuation of a local solution to (1.1.1) which leads to the solution to (1.1.1) defined on the "maximal interval of existence". This process has nothing to do with the analyticity, but is specific to differential equations (one "glues together" appropriate local solutions). Clearly the same "glueing" process can be defined in the complex domain; of course this time it may well lead to a multivalued analytic function. It leads to the same result as the analytic continuation, proviso we do not leave G. Indeed, we already said that all elements of analytic function obtained by the analytic continuation of a local solution inside G are solutions; if two of them are "immediate" continuation of each other, they take the same values at some x, so that they are "glued" during the second process; and if the two local solutions have the same values at some poit x, then they are elements of the same analytic function.

Although theoretically we have a process of analytic continuation, in general one scarcely can say much about the "domain" (to be more precize, Riemann surface) where a given local solution can be continued. But for the linear systems the situation is simple.

Consider first the system

$$\frac{dy}{dx} = A(x)y, \quad y = (y^1, \ldots, y^p)^t, \quad (t \text{ means transposition}) \tag{1.1.2}$$

where the matrix $A(x)$ (i.e. its elements) is (are) holomorphic in the closed disk

$$D = \{x; \; |x - a| \leq r\}$$

(i.e. A is holomorphic in a somewhat bigger open disk). It turns out that any solution $y(x)$ to (1.1.2) is holomorphic in D.

Indeed, let $m = \max_{x \in D} |A(x)|$. As in real domain, write the corresponding integral equation and prove that it has a solution in the whole D. The related estimates for successive approximations

$$y_0(x) = \text{const}, \; y_1(x) = y_0 + \int_a^x A(t)y_0 dt, \ldots, y_n(x) = y_0 + \int_a^x A(t)y_{n-1}(t)dt, \ldots$$

are as follows:

$$|y_n(x) - y_{n-1}(x)| \leq |y_0| \frac{m^n |x - a|^n}{n!}, \quad n = 1, 2 \ldots$$

It is a good training exercise to elaborate another proof, also well-known in the real domain, by glueing together appropriate local solutions, – this idea is called "continuation up to the boundary of the domain". This domain is not D, but D times some big ball in $\mathbb{C}^p$ and the essential point is that as far as y is defined on the linear segment joining a and x,

$$|y(x)| \leq |y_0| e^{m|x-a|},$$

so that the local solutions we are glueing never become "too big".

This allows, first, to show that a solution with arbitrary $y(a)$ is defined in the whole D. We claim also that one can prescribe the value of y at any other point of D and the corresponding solution again will be defined and holomorphic in the whole D. This follows from what we already proved about solutions with prescribed $y(a)$. Take a fundamental system of such solutions and, considering them as columns of Cauchy matrix $Y(x)$, write arbitrary solution as $Y(x)c$ with some constant vector c. And now a solution with prescribed value u at the point b in D will be $y(x) = Y(x)Y^{-1}(b)u$. All this is quite similar to what is well-known in the real domain.

Consider now system (1.1.2) assuming A is holomorphic in

$$S := \bar{\mathbb{C}} \setminus \{a_1, \ldots, a_n\}.$$

Its solutions can be continued along any path in S. Indeed, this path can be covered by a finite number of overlapping disks; and in each of them we have a linear holomorphic system. Thus, we have to "glue" a finite number of local solutions, each being defined in some disk.

Different solutions to different systems with fixed singularities $a_1, \ldots, a_n$ and even different solutions to one system can branch in different ways. But all of them can be lifted to the universal covering surface $\tilde{S}$ of S. For this reason we shall always consider them as holomorphic functions on $\tilde{S}$ (although some of them may well have "a less branching" Riemann surface).

1.2 Introduction

1. Hilbert's 21st problem concerns a certain class of linear ODE's in the complex domain. Let the system

$$\frac{dy}{dx} = A(x)y, \quad y = \begin{pmatrix} y^1 \\ \vdots \\ y^p \end{pmatrix} \tag{1.2.1}$$

have singularities $a_1, \ldots, a_n$; that is, $A(x)$ is holomorphic in $S := \bar{\mathbb{C}} \setminus \{a_1, \ldots, a_n\}$ (where $\bar{\mathbb{C}}$ is the Riemann sphere). The system is called *Fuchsian at a_i* (and a_i is a *Fuchsian singularity* of the system) if $A(x)$ has a pole there of order at most one. The system is *Fuchsian* if it is Fuchsian at all a_i. Let all $a_i \neq \infty$. Then

$$A(x) = \sum_{i=1}^{n} \frac{1}{x - a_i} B_i + B(x),$$

where B is holomorphic on $\mathbb{C}$. We want this system to have no singularity at ∞. First of all let us see when it is Fuchsian at ∞. Rewrite the system in terms of a new independent variable $t = 1/x$. An easy computation shows that

$$\frac{dy}{dt} = (D_1(t) - D_2(t))y, \quad \text{where } D_1 = -\frac{1}{t} \sum_{i=1}^{n} \frac{B_i}{(1 - a_i t)}, \quad D_2 = \frac{1}{t^2} B\left(\frac{1}{t}\right).$$

The matrix function D_1 has a first order pole (or no singularity at all) at $t = 0$. Thus, the system is Fuchsian at $t = 0$ if and only if $\frac{1}{t^2} B(\frac{1}{t})$ has a pole of the first order there. This implies that $B(\infty) = 0$, and since B is holomorphic on $\mathbb{C}$, $B = 0$ everywhere. Hence systems which are Fuchsian on $\bar{\mathbb{C}}$ are just the systems (1.2.1) with A having the form

$$A(x) = \sum_{i=1}^{n} \frac{1}{x - a_i} B_i.$$

(1.2.2)

Such a system has no singularities at ∞ if and only if the residue of D_1 at $t = 0$ is zero, that is

$$\sum_{i=1}^{n} B_i = 0.$$

(1.2.3)

Instead of the vector equation (1.2.1), one considers the matrix equation

$$\frac{dY}{dx} = A(x)Y, \quad Y \text{ is a } (p \times p) \text{ matrix.}$$

(1.2.4)

The columns of Y are p vectors – some solutions $y_1, \ldots, y_p$ to (1.2.1). We shall deal only with the case when they constitute *a fundamental system of solutions*, i.e., they are lineary independent. This means just the invertibility of Y, i.e. $Y \in GL(p, \mathbb{C})$.

Let $p : \tilde{S} \to S$ be the universal covering surface for S. Usually we denote points in S by x and points in $p^{-1}x \subset \tilde{S}$ by $\tilde{x}$. Solutions y and Y are holomorphic functions on $\tilde{S}$, so it is better to write $y(\tilde{x})$, $Y(\tilde{x})$ instead of $y(x)$, $Y(x)$. Let Δ be the group of deck transformations of the covering $p : \tilde{S} \to S$, and let $\sigma, \tau \in \Delta$. Evidently, if y, Y are solutions to (1.2.1), (1.2.4), then so are $y \circ \sigma$, $Y \circ \sigma$. If Y is invertible, then so is $Y \circ \sigma$. However, an invertible solution to (1.2.4) can be obtained from another invertible solution just by multiplying the latter on the right by some constant matrix. Thus

$$Y = (Y \circ \sigma)\chi(\sigma),$$

(1.2.5)

where $\chi : \Delta \to GL(p, \mathbb{C})$ is the so-called *monodromy representation*. It is really a representation, that is,

$$\chi(\sigma\tau) = \chi(\sigma)\chi(\tau).$$

(1.2.6)

Indeed, $Y \circ \tau = [(Y \circ \sigma)\chi(\sigma)] \circ \tau = (Y \circ \sigma \circ \tau)\chi(\sigma)$, so

$$Y = (Y \circ \tau)\chi(\tau) = (Y \circ \sigma \circ \tau)\chi(\sigma)\chi(\tau) = [Y \circ (\sigma\tau)]\chi(\sigma)\chi(\tau),$$

but $Y = [Y \circ (\sigma\tau)]\chi(\sigma\tau)$, so we get (1.2.6). This explains why one prefers (1.2.5) to

$$Y \circ \sigma = Y \chi(\sigma), \tag{1.2.7}$$

which at a first glance may seem more natural. If we choose (1.2.7), then instead of (1.2.6) we get $\chi(\sigma\tau) = \chi(\tau)\chi(\sigma)$, i.e. χ is a so-called anti-representation. It is more convenient to deal with representations.

Instead of $Y = (y_1, \ldots, y_p)$ one can start from another fundamental system of solutions to (1.2.1), i.e. from another invertible solution $\hat{Y}$ of the matrix ODE (1.2.4). Then

$$\hat{Y} = YC \tag{1.2.8}$$

with some constant $C \in GL(p, \mathbb{C})$. Instead of (1.2.5) we get $\hat{Y} = (\hat{Y} \circ \sigma)\hat{\chi}(\sigma)$ with some $\hat{\chi} : \Delta \to GL(p, \mathbb{C})$. So

$$YC = (YC \circ \sigma)\hat{\chi}(\sigma) = (Y \circ \sigma)C\hat{\chi}(\sigma).$$

But $Y = (Y \circ \sigma)\chi(\sigma)$, thus $(Y \circ \sigma)\chi(\sigma)C = (Y \circ \sigma)C\hat{\chi}(\sigma)$. Hence

$$\hat{\chi}(\sigma) = C^{-1}\chi(\sigma)C, \tag{1.2.9}$$

where C is the same for all σ. We see that to a system (1.2.1) there corresponds a class of mutually conjugate representations $\Delta \to GL(p, \mathbb{C})$. We shall call this class simply *monodromy*. For any representation χ_1 belonging to this class there exists an invertible matrix solution Y_1 to (1.2.4) such that $Y_1 = (Y_1 \circ \sigma)\chi_1(\sigma)$. (If $\chi_1 = C^{-1}\chi C$, take $Y_1 = YC$).

Consider the space of all solutions $y = y(\tilde{x})$ to (1.2.1). This is a p-dimensional vector space X. For $y \in X$, $\sigma \in \Delta$ let $\sigma^* y := y \circ \sigma^{-1}$, i.e., $(\sigma^* y)(\tilde{x}) = y(\sigma^{-1}\tilde{x})$. Clearly $\sigma^* y$ is also a solution to (1.2.1), so we obtain a map $\sigma^* : X \to X$ which is an invertible linear transformation. This defines a map

$$\Delta \to GL(X), \quad \sigma \mapsto \sigma^*.$$

It is easy to check that $(\sigma\tau)^* = \sigma^*\tau^*$, i.e., this map is a homomorphism. (If we defined σ^* as $\sigma^* y := y \circ \sigma$, then $\sigma \mapsto \sigma^*$ would be an anti-homomorphism). After choosing a basis $y_1, \ldots, y_p$ in X, we can identify X and $GL(X)$ with the more concrete objects $\mathbb{C}^p$ and $GL(p, \mathbb{C})$. This basis defines a map

$$\tilde{S} \to GL(p, \mathbb{C}), \quad \tilde{x} \mapsto Y(\tilde{x}) = (y_1(\tilde{x}), \ldots y_p(\tilde{x}))$$

(y_j are the columns of the matrix Y). Clearly $Y(\tilde{x})$ is a solution to the matrix equation (1.2.4) and $(\sigma^* y_1, \ldots, \sigma^* y_p) = Y \circ \sigma^{-1}$. It follows that the monodromy matrix $\chi(\sigma)$ is just the matrix describing the linear map $\sigma^* : X \to X$ with respect to the basis $y_1, \ldots, y_p$ in X.

2. Hilbert's 21st problem is stated as follows ([Hi1]): *Prove that there always exists a Fuchsian system with given singularities and a given monodromy.* Hilbert himself said "prove", but it would be more careful to say "inquire whether..." This is a distinctly formulated problem which has to be answered "yes" or "no" (whereas some of Hilbert's problems are formulated not so distinctly, e.g.: "develop the calculus of variations along such and such lines").

Literally, Hilbert said "equation", not "system". (For equations one also has a notion of Fuchsian equations, see (1.2.12). The monodromy for the pth order linear ODE is just the same as for the pth order system describing the behavior of the vectors $\left(y, \frac{dy}{dx}, \ldots, \frac{d^{p-1}y}{dx^{p-1}} \right)$, where y satisfies the equation). Does one have to understand this as "a system of equations" (as we often do in conversations and even in the titles of textbooks)? We think that the answer is "yes", because it was already known at the time that for equations the same problem has a negative answer. It is very easy to verify that a Fuchsian equation of pth order with singularities $a_1, \ldots, a_n$ contains fewer parameters than the set of classes of conjugate representations $\Delta \to GL(p, \mathbb{C})$. (This goes back to Poincaré [Poi]), who calculated the difference between these two numbers of parameters, see Chapter 7). So in general it is impossible to construct a Fuchsian equation without an appearance of additional singularities. But the clear and accurate statement of Hilbert's 21st problem does not allow such a possibility.

In mathematical literature Hilbert's 21st problem is often called the Riemann-Hilbert problem, although Riemann never spoke exactly of something like it. This was well-known: Klein in his "Lectures on the development of the mathematics in 19th century" [Kl] said that "Riemann speaks in such a careless way as if existence of functions $y_1, \ldots, y_p$ (having the given singularities and monodromy) is self-evident and one has only to study their properties". However, Hilbert mentioned that "presumably Riemann was thinking on this problem", and Röhrl [Rö] made a final step in this mythological direction and distinctly attributed Hilbert's 21st problem to Riemann. As well as the majority of the mathematicians who have dealt with the problem we prefer to say that the Riemann-Hilbert problem (Hilbert's 21st problem) is close to the sphere of Riemann's ideas and it has arisen in the course of research stimulated by him.

For a number of years people thought that Hilbert's 21st problem was completely solved by Plemelj [Pl] in 1908. Only recently it was realized that there was a gap in his proof (for the first time this was observed by T.Kohn [Koh] and V.I.Arnold,

Yu.S. Il'yashenko [ArIl]). It turned out that Plemelj obtained a positive answer to a problem similar to Hilbert's 21st problem but concerning so-called regular systems instead of Fuchsian ones. Here is the definition of them.

Let (1.2.1) be a system with singularities $a_1, \ldots, a_n$. It is called *regular at* a_i (and a_i is *a regular singularity* for this system), if any of its solutions has at most polynomial growth as $x \to a_i$. ("Polynomial" means "polynomial in $1/|x - a_i|$"). This has to be stated more carefully, as it is clear that if y branches logarithmically near a_i, then one can get any growth of y as x tends to a_i along an appropriate spiral: each turn gives a constant "increase" to y. One must demand that $x \to a_i$ in an "honest" way, remaining inside some sector Σ having vertex at a_i. Here is the precise definition: *for any such sector* Σ, *for any "covering" sector* $\tilde{\Sigma}$ *on* $\tilde{S}$ *and for any solution* y, *the restriction* $y|\tilde{\Sigma}$ *has at most polynomial growth as* $x \to a_i$ *remaining in* Σ. It is sufficient to demand this for only p linearly independent solutions, or equivalently, for an invertible matrix solution Y to (1.2.4). In view of (1.2.5) one needs only take several sectors $\tilde{\Sigma}_h \subset \tilde{S}$ such that $p\left(\cup\tilde{\Sigma}_h\right)$ is a disk centered at a_i. It follows that there exists $\lambda \in \mathbb{R}$ such that for any $\Sigma, \tilde{\Sigma}$ and y as before,

$$\frac{y(\tilde{x})}{|x - a_i|^\lambda} \to 0 \text{ as } x \to a_i \quad \left(p\tilde{x} = x \in \Sigma, \tilde{x} \in \tilde{\Sigma}\right). \tag{1.2.10}$$

The system is called *regular* if it is regular at all a_i. Any Fuchsian system is regular (see [Ha] Hartman's textbook on ODE's for a very short proof due to G. Birkhoff), but a regular system need not be Fuchsian (Plemelj was able to find systems in a broader class than required by Hilbert).

Of course, misunderstandings are common for human's activity, but it is not so common that a misunderstanding in mathematics retains for more than 70 years. Perhaps this is explained by the following circumstances. First, the notions of "Fuchsianity" and "regularity" are defined in some way also for the pth order scalar linear ODE in the complex domain. In this case these notions, being different in appearance, turn out to be equivalent. This is not the case for the systems – it is well-known and trivial that for them the two notions under consideration are really different. But now the second circumstance appears: locally it is easy to modify a regular system so that one gets a Fuchsian system with the same singular point and monodromy. Perhaps all this supports the unconcious feeling that also globally one can modify the regular system so that one gets a Fuchsian system with the same singularities and monodromy. Now we know that this feeling is wrong, although it is easy to pass to a Fuchsian system with the same monodromy if we admit the extra singularities. (The latter are called *apparent singularities*, although they are true singularities in the sense that they are poles of the coefficients. They are apparent only in the sense that the solutions do not branch there). However, this feeling is not entirely deceptive: as we already mentioned, almost all positive results to Hilbert's 21st problem for Fuchsian systems are obtained by perfoming some appropriate modification of a regular system provided by the Plemelj theorem.

Although Hilbert spoke of Fuchsian systems, one may ask whether he could have in mind regular ones. Here again arises the problem of interpretations etc. which we consider as a difficult one. But if he could have this in mind, it means that there was some ambiguity with the term "Fuchsian" at that time. Then one should consider Hilbert's 21st problem as consisting of two parts – one for Fuchsian, another for regular systems. Whether it is justified historically or not, such a point of view on this problem is quite reasonable as really there are two problems.

Some particular cases of Hilbert's 21st problem were more or less solved (always positively) before Plemelj (some of them – even before Hilbert published his list of 23 problems). References are given in Hilbert's text [Hi1] devoted to these problems and in Röhrl's paper [Rö]; also Klein refers to Hilb's (not Hilbert's) article in the German Math. Encycl. However, nowadays one must check whether these results concerned Fuchsian or regular systems and whether they were proved at all; this explaines why we said "more or less". Hilbert himself published a paper [Hi2]. Klein refers to Hilbert's solution of the problem in the general case, but Röhrl attributes to him only the positive solution for the case of two equations and any number of singularities. At any case, Hilbert's paper has a reputation of involved etc. and it seems that after Plemelj published a much more lucid paper nobody was interested in the careful analysis of Hilbert's arguments.

Plemelj used the theory of singular integral equations which he developed especially for this purpose. (Perhaps this was the first success both in developing and applying this theory). In 1957 Röhrl ([Rö]) published another approach to the same problem using some arguments from the theory of Riemann surfaces and the algebraic geometry. There are some improvements of his approach. Primarily they go back to Deligne [Del] [1]; also several remarks are due to [Bo4] and [An]. Taking into account all this improvements, Röhrl's approach can be considered as an elementary one, with an essential exception: one has to use a nontrivial theorem due to Birkhoff and Grothendieck (or another statement which is perhaps one slightly weaker; both will be stated below). Birkhoff proved this theorem using singular integral equations, while Grothendieck used algebraic geometry; so it may seem that the reduction of our problem to this theorem only moves more difficult arguments to another place.

---

[1] He considered an analogous problem for a system of Pfaffian differential equations with several independent variables. This made a more geometric point of view more or less unavoidable. But the same point of view turns out to be useful in our case. (Of course here it becomes simpler. E.g., we need not mention Deligne's "flat connections" explicitly, although the "branching cross section" Z below is a simple manifestation of the same idea).

But now there exists a short and elementary proof of the Birkhoff-Grothendieck's theorem. It was published by Leiterer [Lei].

The goal of the third Chapter of our book is to give a complete proof of Plemelj's theorem, following Röhrl's approach, using the improvements mentioned above and including a slightly modified version of the proof of Birkhoff-Grothendieck's theorem sketched by Leiterer. In our opinion, both the reduction of the problem to this theorem and the proof of the latter use nontrivial ideas (although they are elementary); however, there is a difference in style between this two parts of the argument. First part looks like "abstract nonsense", so it is almost trivial in appearance (but it is quite nontrivial that this can be made trivial!). The second part seems to be nontrivial both in appearance and in essence.

In 1989 it was found an unexpected negative solution to Hilbert's 21st problem in [Bo1], [Bo2]. It is explained in Chapter 2. This result changes our point of view on these questions and increases the value of various partial positive results, such as the following:

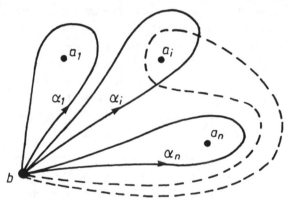

1. (Plemelj, [Pl]). *Fix $b \in S$ and let $\alpha_1, \ldots, \alpha_n$ be loops at b such that α_i "goes around" a_i without "going around" any $a_j \neq a_i$. (Such a system of loops is not uniquely determined – see the dashed line in the figure, but anyone of them will do). If at least one of the matrices $\chi(\alpha_1), \ldots, \chi(\alpha_n)$ is semisimple (diagonalizable), then the answer is positive.*

2. (Lappo-Danilevskii, [LD]. 1920-s). *If all $\chi(\alpha_i)$ are sufficiently close to the identity matrix I, then the answer is positive.*

3. (Dekkers, [Dek], 1979). *In the case $p = 2$ the answer is positive (independently of n).*[2]

4. (Kostov, [Ko1], [Ko2], Bolibruch, [Bo5], [Bo6]). *If the representation χ is irreducible, then the answer is positive.*

5. (Bolibruch, [Bo2]). *If $p = 3$ then there is a complete answer whether the problem has a positive or negative solution for a given χ.* (See Chapter 6 of the present book).

It is worth mentioning once more that the proofs of several positive results about the Hilbert 21st problem for Fuchsian systems begin by referring to Plemelj's theorem on regular systems; after this one modifies the regular system provided by this theorem in order to obtain a Fuchsian system with the same a_i, χ.

The goal of the second Chapter of this book is to explain the first negative result. Not only does it provide an answer to Hilbert's 21st problem but also serves as an introduction to other results.

The first negative result concerns the case $p = 3, n = 4$. (It is the first case when the known positive results do not apply – and now we understand why).

It was found out that this "counterexample" has the following property: if one perturbs the singular points a_i then the answer to Hilbert's 21st problem with the same monodromy can become positive ([Bo2]). Thus, in the "counterexample" the monodromy must be somehow tied to the singular points. In the theory of ODE's it does not seem unnatural to tie these things by means of a differential equation. Indeed, this is the case for the "first counterexample".

Some later in [Bo4] there were obtained new series of representations, that give a negative solution to Hilbert's 21st problem and are already stable under perturbations of singular points. These series and some other new results concerning Hilbert's 21st problem are presented in Chapter 5.

3. In Chapter 7 we consider a connection between Fuchsian systems and Fuchsian equations on the Riemann sphere. Equation

$$y^{(p)} + q_1(x)y^{(p-1)} + \ldots + q_p(x)y = 0 \qquad (1.2.11)$$

is called *Fuchsian at a point* a, if its coefficients $q_1(x), \ldots, q_p(x)$ are holomorphic in some punctured neighborhood of this point and

---

[2]Lappo-Danilevskii and Dekkers did not pretend to solve the Riemann-Hilbert problem (that time there was the opinion that this problem was solved by Plemelj), but the results formulated above follow immediately from their results.

$$q_i(x) = \frac{r_i(x)}{(x-a)^i}, \quad i = 1, \ldots, p \tag{1.2.12}$$

where $r_1(x), \ldots, r_p(x)$ are functions holomorphic at a. All solutions of a Fuchsian system have at most a polynomial growth at a_i, so the Fuchsian point is regular. This has to be made more precise in the same way as for the systems. It tuns out that *(1.2.11) is regular at the point a if and only if the system describing the behavior of the vector $(y^1, \ldots, y^p) = \left(y, \frac{dy}{dx}, \ldots, \frac{d^{p-1}y}{dx^{p-1}}\right)$, i.e., the system*

$$
\begin{aligned}
\frac{dy^1}{dx} &= y^2 \\
\frac{dy^2}{dx} &= y^3 \\
\cdots \quad & \cdots\cdots\cdots\cdots \\
\frac{dy^p}{dx} &= -q_p(x)y^1 - \ldots - q_1(x)y^p,
\end{aligned}
\tag{1.2.13}
$$

is regular at a. It is well-known that *(1.2.11) is regular at $x = a$ if and only if it is Fuchsian at $x = a$* (see [Ha]).

For the systems the analogous statement is not valid. Equation

$$\frac{d^2y}{dx^2} + \frac{1}{x}\frac{dy}{dy} + \frac{1}{x^2}y = 0$$

is Fuchsian at $x = 0$, hence the corresponding system (1.2.13) is regular there, but one of its coefficients has a pole of the 2nd order.

Nevertheless there are ways to transform (1.2.11) to a system Fuchsian at a. For this purpose we can replace transformation (1.2.13) by the following one:

$$
\begin{aligned}
y &= z^1 \\
(x-a)\frac{dy}{dx} &= z^2 \\
\cdots\cdots & \cdots\cdots\cdots \\
(x-a)^{p-1}\frac{d^{p-1}y}{dx^{p-1}} &= z^p.
\end{aligned}
\tag{1.2.14}
$$

Under such a transformation we get

$$(x-a)\frac{dz^1}{dx} = z^2$$

$$(x-a)\frac{dz^2}{dx} = z^2 + z^3 \qquad (1.2.15)$$

$$\cdots \qquad \cdots$$

$$(x-a)\frac{dz^p}{dx} = (p-1)z^p - \sum_{k=1}^{p}(z-a)^{p-(k-1)}q_{p-k+1}(x)z^k.$$

Due to (1.2.12), (1.2.15) we obtain that the vector z satisfies the system (1.2.1) with the matrix

$$A(x) = \frac{1}{x}\begin{pmatrix} 0 & 1 & 0 & 0 & \cdots & 0 \\ 0 & 1 & 1 & 0 & \cdots & 0 \\ 0 & 0 & 2 & 1 & \cdots & 0 \\ \cdots & \cdots & \cdots & \cdots & \cdots & \cdots \\ -r_p & -r_{p-1} & \cdots & \cdots & \cdots & (p-1)-r_1 \end{pmatrix},$$

thus the system is Fuchsian at a.

Equation (1.2.11) is called *Fuchsian* (on the whole Riemann sphere) if all $q_i(x)$ are holomorphic on $S = \bar{\mathbb{C}} \setminus \{a_1, \ldots, a_n\}$ and (1.2.11) is Fuchsian at points $a_1, \ldots, a_n$.

In Chapter 7 we prove that any Fuchsian equation (1.2.11) can be transformed to a Fuchsian linear system with the same singular points and a monodromy group on the whole Riemann sphere without appearance of new singularities.

Here we also estimate the number of so-called "apparent" singularities of a Fuchsian linear differential equation of p-th order. (These singularities appear under attempts to construct a Fuchsian linear differential equation of the p-th order with a given monodromy χ).

4. There are different modifications and generalizations of the classical Riemann-Hilbert problem. The analogous problem over Riemann surfaces were considered by Röhrl [Rö] (see also [Fö]). Some modifications of the problem over the Riemann sphere were considered by G.Birkhoff [Bi] and Il'yashenko [ArII].

Röhrl's and Deligne's papers [Del] gave rise to the setting and investigation of the multidimentional Riemann-Hilbert problem in [Ge], [Go1], [Lek], [Su], [Ki], [Hai].

Hilbert's 21st problem and its analogous have many applications in various areas of mathematics and physics, which are not discussed in our book. Information on these subjects can be found in [Ka], [SJM], [Go2].

2 Counterexample to Hilbert's 21st problem

2.1 The first counterexample

Consider the system (1.2.1) with

$$A(x) = \frac{1}{x^2} \begin{pmatrix} 0 & 1 & 0 \\ 0 & x & 0 \\ 0 & 0 & -x \end{pmatrix} + \frac{1}{6(x+1)} \begin{pmatrix} 0 & 6 & 0 \\ 0 & -1 & 1 \\ 0 & -1 & 1 \end{pmatrix} + \qquad (2.1.1)$$

$$+ \frac{1}{2(x-1)} \begin{pmatrix} 0 & 0 & 2 \\ 0 & -1 & -1 \\ 0 & 1 & 1 \end{pmatrix} + \frac{1}{3\left(x - \frac{1}{2}\right)} \begin{pmatrix} 0 & -3 & -3 \\ 0 & -1 & 1 \\ 0 & -1 & 1 \end{pmatrix}.$$

It is singular at $a_0 = 0$, $a_1 = -1$, $a_2 = 1$, $a_3 = \frac{1}{2}$. Later (in Section 2.4) we will check that there is no singularity at ∞. The points a_1, a_2, a_3 are Fuchsian singularities, however a_0 is not Fuchsian, but a pole of order 2. Thus, the system is not Fuchsian, but we shall see that it is regular.

Our system has some monodromy. We shall prove the following assertion. *There exists no Fuchsian system with the same singularities and monodromy.* (This statement will be referred to as the "Assertion".) In this example the monodromy is given implicitly. Here are some comments on this fact.

The Assertion is very sensitive to the given data (a_i and χ) – one type of sensitivity was described in the penultimate paragraph of item 2 of Section 1.2 (sensitivity to a_i), another is evident from Plemelj's positive result (1) above. This second type of sensitivity does not depend on p. The first type does depend on p (at least sometimes): for $p = 4, n \geq 3$ the second author was able to find (explicitly) a monodromy which cannot occur in a Fuchsian system whatever the $a_i's$ are (see Chapter 5).

Now about the method of proof of the Assertion. The matrix (2.1.1) has the form

$$A(x) = \begin{pmatrix} 0 & a_{12}(x) & a_{13}(x) \\ 0 & & \\ & & B(x) \\ 0 & & \end{pmatrix} \qquad (2.1.2)$$

where

$$a_{12}(x) = \frac{1}{x^2} + \frac{1}{x+1} - \frac{1}{x-\frac{1}{2}}; \; a_{13}(x) = \frac{1}{x-1} - \frac{1}{x-\frac{1}{2}}; \qquad (2.1.3)$$

$$B(x) = \frac{1}{x} \begin{pmatrix} 1 & 0 \\ 0 & -1 \end{pmatrix} + \frac{1}{6(x+1)} \begin{pmatrix} -1 & 1 \\ -1 & 1 \end{pmatrix} + \qquad (2.1.4)$$

$$+ \frac{1}{2(x-1)} \begin{pmatrix} -1 & -1 \\ 1 & 1 \end{pmatrix} + \frac{1}{3(x-\frac{1}{2})} \begin{pmatrix} -1 & 1 \\ -1 & 1 \end{pmatrix}.$$

For the system (1.2.1), (2.1.2)

$$\frac{dy^1}{dx} = a_{12}(x)y^2 + a_{13}(x)y^3, \qquad (2.1.5)$$

where y^2 and y^3 satisfy the system

$$\frac{dy}{dx} = B(x)y, \quad y \in \mathbb{C}^2 \qquad (2.1.6)$$

with B as in (2.1.4). When studying this system in its own right, we shall write y^1, y^2 instead of y^2, y^3. (This will not be much of an inconvenience.) Clearly the properties of (2.1.6), (2.1.4) are important for the study of (1.2.1), (2.1.1), so Section 2.3 will be devoted to them.

Assume that our Assertion is false, i.e., there exists a Fuchsian system

$$\frac{dy}{dx} = C(x)y, \; y \in \mathbb{C}^3 \qquad (2.1.7)$$

with the same singularities and monodromy as (1.2.1), (2.1.1). The latter already indicates some similarity of these systems. In Section 2.4 we shall transform (2.1.7) in order to increase this similarity. Afterwards, we shall be able "to pick out" from the modified "C-system" a second order Fuchsian quotient system

$$\frac{dv}{dx} = F(x)v, \; v \in \mathbb{C}^2 \qquad\qquad (2.1.8)$$

having the same singularities and monodromy as (2.1.6), (2.1.4) and satisfying some extra conditions. The investigation theorem in Section 2.3 will show that such a system cannot exist.

We shall rephrase the very end of previous paragraph. Consider the "problem": [P] *"Does there exist a second order Fuchsian system having the same singularities and monodromy as (2.1.6), (2.1.4)?"* [P] is trivially "yes": we need not even appeal to Dekkers ((3) above), as (2.1.6), (2.1.4) is itself Fuchsian. However, if we impose some additional requirements on the system we are looking for, the answer may be "no". Earlier we said essentially that the falseness of the Assertion implies a positive answer to the problem [P+ some extra requirement]. Thus, we see that a negative answer to Hilbert's 21st problem for $p = 3$ depends upon a negative answer to a related problem for $p = 2$. Perhaps this auxialiary problem in its own right seems somewhat unnatural, at least less natural, but that does not matter (and depends more on the experience than on the taste).

Just to state the auxialiary problem one needs a new local theory to supplement the well-known theory which goes back to Poincaré and can be found in such textbooks as those by [CoLe] or [Ha]. (Needless to say one needs the new theory in order to investigate the auxialiary problem and its relationship to the Assertion.) This new theory is due to Levelt (1961) [Le]. Section 2.2 is devoted to it.

2.2   Local theory

First we recall the old theory. We shall consider the system (1.2.1) near an isolated singular point, say 0. Let U be a small disk with center 0, $U^* = U \setminus 0$, $p : \tilde{U}^* \to U^*$ be the universal covering of U^*. A(x) is holomorphic in U^*, solutions y (to (1.2.1)) and Y (to (1.2.4)) are holomorphic in $\tilde{U}^*$.

The group of deck transformations Δ is now an infinite cyclic group generated by a deck transformation σ which corresponds to one trip around 0 counterclockwise. Clearly $\ln \tilde{x}$ is a holomorphic function in $\tilde{U}^*$ and $\ln(\sigma \tilde{x}) = \ln \tilde{x} + 2\pi i$. Let $G = \chi(\sigma^{-1})$ so that

$$Y(\sigma \tilde{x}) = Y(\tilde{x})G \qquad\qquad (2.2.1)$$

(which is similar to (1.2.7) – for a cyclic group we do not bother with the difference between representations and anti-representations, so there are no objections to

(1.2.7)). Let $E = \frac{1}{2\pi i} \ln G$ (logarithm in the sense of the matrix theory), so that if λ^j are eigenvalues of G and μ^j of E, then $\mu^j = \frac{1}{2\pi i} \ln \lambda^j$. Denote $\rho^j = Re\mu^j$ and normalize the choice of ln demanding that

$$0 \le \rho^j < 1 \qquad (2.2.2)$$

(this is not necessary for the Poincaré arguments, but we shall use it later). Introduce the function $\tilde{x}^E := e^{E \ln \tilde{x}}$ (holomorphic on $\check{U}^*$);

$$(\sigma\tilde{x})^E = e^{E(\ln \tilde{x} + 2\pi i)} = \tilde{x}^E G.$$

Then (cf. (2.2.1))

$$Y(\sigma\tilde{x})(\sigma\tilde{x})^{-E} = Y(\tilde{x})GG^{-1}\tilde{x}^{-E} = Y(\tilde{x})\tilde{x}^{-E}.$$

Hence, $Y(\tilde{x})\tilde{x}^{-E}$ can be considered as a single-valued holomorphic function on U^*; denote it by $Z(x)$. We arrive at Poincaré result claiming that any (invertible) solution Y to (1.2.4) can be represented as follows:

$$Y(\tilde{x}) = Z(x)\tilde{x}^E, \qquad (2.2.3)$$

where Z is holomorphic on U^*. Recall that G in (2.2.1) (and hence E here) depends on Y; for $\tilde{Y} = YC$ one must replace G by $\hat{G} = C^{-1}GC$ (Cf. (1.2.8),(1.2.9)).

Now we turn to Levelt's theory. It concerns only regular systems. For a scalar or vector function holomorphic on $\check{U}^*$ and having only polynomial growth when $x \to 0$ (cf. (1.2.10) and the discussion preceding it) define

$$\varphi(y) : = \left[\sup\left\{\lambda; \frac{y(\tilde{x})}{|x|^\lambda} \to 0 \text{ as } x \to \infty\right\}\right] =$$
$$= \max\left\{k \in \mathbb{Z}; \forall \lambda < k \quad \frac{y(\tilde{x})}{|x|^\lambda} \to 0 \text{ as } x \to 0\right\} \qquad (2.2.4)$$
$$\varphi(0) = \infty.$$

(Here $[\ldots]$ denotes the entire part. As regards to the statements of the type "$\ldots \to 0$ as $x \to 0$", they are subject to the same provisos as before). For example:

$$\varphi\left(\frac{1}{x}\ln\tilde{x}\right) = \varphi\left(\frac{1}{\sqrt{\tilde{x}}}\right) = -1$$

Evidently,

$$\text{if } y = \begin{pmatrix} y^1 \\ \vdots \\ y^p \end{pmatrix}, \text{ then } \varphi(y) = \min \varphi(y^j) \tag{2.2.5}$$

also

$$\varphi(y \circ \sigma) = \varphi(y). \tag{2.2.6}$$

Indeed, our

$$\text{,,} \frac{y(\tilde{x})}{|x|^\lambda} \to 0 \text{ as } x \to 0\text{''} \tag{2.2.7}$$

means that whenever Σ, $\tilde{\Sigma}$ are as in the text preceding (1.2.10),

$$\frac{\left(y | \tilde{\Sigma} \right)(\tilde{x})}{|x|^\lambda} \to 0 \text{ as } x \to 0, \ x \in \Sigma, \ \tilde{x} \in \tilde{\Sigma}, \ p\tilde{x} = x.$$

But this is equivalent to

$$\frac{\left((y \circ \sigma) | \sigma^{-1}\tilde{\Sigma} \right)(\tilde{x})}{|x|^\lambda} \to 0 \text{ as } x \to 0, \ x \in \Sigma, \ \tilde{x} \in \sigma^{-1}\tilde{\Sigma}, \ p\tilde{x} = x,$$

since we can replace $\tilde{x} \in \sigma^{-1}\tilde{\Sigma}$ in the latter formula by $\sigma^{-1}\tilde{x}$, $\tilde{x} \in \tilde{\Sigma}$ (still $p\tilde{x} = x$), and then in the numerators in both formulas we shall have the same function

$$\Sigma \to \mathbb{C}, \ x \to y(\tilde{x}) = y(\sigma\sigma^{-1}\tilde{x}), \ \tilde{x} \in \tilde{\Sigma}, \ p\tilde{x} = x.$$

As $\tilde{\Sigma}$ runs over all sectors covering Σ, so does $\sigma^{-1}\tilde{\Sigma}$. Hence (2.2.7) is equivalent to

$$\text{,,} \frac{(y \circ \sigma)(\tilde{x})}{|x|^\lambda} \to 0 \text{ as } x \to 0\text{''}$$

(quotes indicating the same provisos as in (2.2.7)), and the set of λ in (2.2.4) is the same for y and $y \circ \sigma$.)

Solutions y to (1.2.1) (recall that they are some vector functions on $\tilde{U}^*$) constitute a vector space X isomorphic to $\mathbb{C}^p$. Restricted to X, φ is a map $\varphi : X \to \mathbb{Z}$ having the following properties:

$$\varphi(\lambda y) = \varphi(y), \text{ if } \lambda \in \mathbb{C} \setminus 0; \quad \varphi(0) = \infty; \qquad (2.2.8)$$
$$\varphi(y_1 + y_2) \geq \min(\varphi(y_1), \varphi(y_2)), \text{ with equality if } \varphi(y_1) \neq \varphi(y_2).$$

In algebraic terminology this can be expressed by the words: "φ is a nonarchimedian normalization on X over the trivial normalization on $\mathbb{C}$". The appearance of the root "norm" is explained as follows. If we define

$$\||\lambda\|| = 1 \text{ when } \lambda \in \mathbb{C} \setminus 0, \ \||0\|| = 0,$$
$$\||y\|| = a^{-\varphi(y)} \text{ for } y \in X \text{ (using a fixed } a > 1),$$

then $\|| \cdot \||$ satisfies the standard properties of the norms:

$$\||\lambda y\|| = \||\lambda\|| \cdot \||y\||$$
$$\||y_1 + y_2\|| \leq \||y_1\|| + \||y_2\||.$$

(Moreover: $\||y_1 + y_2\|| \leq \min(\||y_1\||, \||y_2\||)$, with equality when $\||y_1\|| \neq \||y_2\||$. Of course this is not standard, but a peculiar property due to the nonarchimedian character of φ). However, when dealing with nonarchimedian norms, people usually work with such functions as φ, without appealing to $\|| \cdot \||$.

Another example of a normalization on a finite dimensional vector space X is given by the Lyapunov characteristic numbers (exponents). In this case, X is the space of solutions $x(t)$ to a linear system $\frac{dx}{dt} = A(t)x$ on $[0, \infty)$ with appropriate restrictions on A. The characteristic number of $x \in X$ is

$$\chi(x) := \overline{\lim}_{t \to \infty} \frac{1}{t} \ln |x(t)|, \ \chi(0) := -\infty.$$

Then $\varphi = -\chi$ has the properties (2.2.8). In contrast to Levelts's case, this φ takes its values in $\mathbb{R} \cup \infty$. Properties of χ are well-known, and it is more or less well-known that many of them are due to the fact that $\varphi = -\chi$ satisfies (2.2.8). (Of course one does not go from χ to $\varphi = -\chi$, but just rewrites (2.2.8) in terms of χ. Incidentally, Lyapunov himself defined χ as the negative of the definition given above). Due to Levelt further we shall use term *valuation* for φ.

In any case, one can prove that φ defines a filtration of X (a strictly increasing sequence of vector subspaces)

$$0 = X^0 \subset X^1 \subset X^2 \subset \cdots \subset X^h = X \qquad (2.2.9)$$

such that φ is constant on $X^j \setminus X^{j-1}$ and if $\psi^j = \varphi(X^j \setminus X^{j-1})$ then $\psi^1 > \psi^2 > \cdots > \psi^h$. Let $k_j := \dim X^j - \dim X^{j-1}$. We say that φ takes the value ψ^j with the multiplicity k_j, or that ψ^j has multiplicity k_j. We shall use also the notation

$$\varphi^1 = \cdots = \varphi^{k_1} = \psi^1,$$
$$\varphi^{k_1+1} = \cdots = \varphi^{k_1+k_2} = \psi^2,$$
$$\cdots \cdots \cdots$$
$$\varphi^{k_1+\cdots+k_{h-1}+1} = \cdots = \varphi^{k_1+\cdots+k_{h-1}+k_h} = \psi^h.$$

Note that

$$\varphi^1 \geq \varphi^2 \geq \cdots \geq \varphi^p. \qquad (2.2.10)$$

There exists a basis $y_1, \ldots, y_p$ in X such that $\varphi(y_j) = \varphi^j$. We take k_1 linearly independent vectors in X^1, then add to this collection k_2 vectors in X^2 which are linearly independent mod X^1, and so on.

Until now we have used only (2.2.8). Recall now that our y's are solutions to (1.2.1) – some vector functions on $\tilde{U}^*$ where the deck transformation σ acts. If y is a solution to (1.2.1), then $y \circ \sigma$ is also a solution (again defined on $\tilde{U}^*$), so we get a linear transformation[1]

$$\sigma^* : X \to X \quad \sigma^* y = y \circ \sigma.$$

It preserves the filtration (2.2.9) (cf. (2.2.6). Consider the induced transformation σ_j^* on the jth factorspace and take a basis $\bar{y}_{1j}, \ldots \bar{y}_{k_j j}$ in this space that σ_j^* has an upper-triangular(say, Jordan) matrix representation in this basis. Each $\bar{y}_{lj}$ is some coset $y_{lj} + X^{j-1}$ where y_{lj} is any representative of this coset. Take the following basis in X:

$$y_{11}, \ldots, y_{k_1 1}, y_{12}, \ldots, y_{k_2 2}, \ldots, y_{1h}, \ldots, y_{k_h h}.$$

---

[1]In Chapter 1 we defined $\sigma^* y$ slightly differently, $\sigma^* = y \circ \sigma^{-1}$. As we mentioned already, in the local situation one has cyclic group of deck transformations and needs not bother with the difference between representations and anti-representations. The matrix G in (2.2.1) is just the matrix describing σ^* with respect to the basis whose vectors are columns of Y.

Denote these vectors (in this order) by $y_1, \ldots, y_p$. Such a choice of basis is a particular case of the choice considered in the previous paragraph. We conclude that $\varphi(y_j) = \varphi^j$ and σ^* has an upper-triangular representation in this basis. To check the latter, argue as follows: if $y_j = y_{lm}, 1 \le l \le k_m, 1 \le m \le h$, then

$$\sigma_m^* \bar{y}_{lm} \in \sum_{q=1}^{l} \mathbb{C} \bar{y}_{qm} \text{ in } X^m / X^{m-1}$$

and in X

$$\sigma^* y_j = \sigma^* y_{lm} \in \sum_{q=1}^{l} \mathbb{C} y_{qm} + X^{m-1} =$$

$$= \sum_{q=1}^{l} \mathbb{C} y_{qm} + \sum_{r=1}^{m-1} \sum_{q=1}^{k_r} \mathbb{C} y_{qr} = \sum_{s=1}^{j} \mathbb{C} y_s.$$

We shall call such a basis *a Levelt's basis* or *a Levelt's fundamental system of solutions to (1.2.1)* (although Levelt himself used another name). It is clear from the construction that it is not unique, i.e., there is some freedom in choosing it, but in general only some bases are Levelt's bases. (Note that a Levelt's basis, by definition, is an ordered system of vectors; in general the same vectors taken in another order will not constitute a Levelt's basis). A matrix $Y = (y_1, \ldots, y_p)$ whose columns constitute a Levelt's basis we shall call *a Levelt's matrix* or *a Levelt's (matrix) solution (to (1.2.4))*. Now we shall explain that for a Levelt's matrix Poincaré representation (2.2.3) can be improved.

Let us note first if one uses some basis $(y_1, \ldots, y_p)$ in X (not necessarily a Levelt's basis), the matrix representation of σ^* in this basis is just given by the monodromy matrix G related to $Y = (y_1, \ldots, y_p)$ according to (2.2.1). Indeed, what does it mean that some matrix, say H, is the matrix representation of σ^* in the basis $(y_1, \ldots, y_p)$? This means the following. Take a vector z having coordinates $\zeta = (\zeta^1, \ldots, \zeta^p)$ in this basis. Then $\sigma^* z$ has coordinates $H\zeta$ (writing ζ as a column vector). Now $z = \sum_{j=1}^{p} \zeta^j y_j$, which can also be written as $z = Y\zeta$. Clearly, $\sigma^* z = \sum_{j=1}^{p} \zeta^j(y_j \circ \sigma) = (Y \circ \sigma)\zeta = YG\zeta$. But this means that $\sigma^* z$ has coordinates $G\zeta$ in the basis $(y_1, \ldots, y_p)$, i.e., G is actually the matrix representation discussed.

We conclude that if $(y_1, \ldots, y_p)$ is a Levelt's basis then G is an upper-triangular matrix. Hence, so are E and $e^{tE} (t \in \mathbb{C}), \tilde{x}^E = e^{\ln \tilde{x} E}$. Note that $X(t) := e^{tE}$ satisfies the system of ordinary differential equations $\frac{dX}{dt} = EX$ having constant and upper-triangular coefficient matrix, with $X(0) = I$ (the identity matrix). Writing $X(t) = (x_{jk}(t))$, it follows easily that $x_{jj}(t) = e^{\mu^j t}$ (recall that μ^j are the eigenvalues of E)

and each $x_{jk}(t)$ with $j < k$ is a sum of products of some polynomials in t and some exponential functions $e^{\mu' t}$ (whereas $x_{jk} = 0$ for $j > k$). Denote the coefficients of $\tilde{x}^E$ by $\alpha_{kj}(\tilde{x})$. Then

$$\varphi(\alpha_{kj}) \geq 0 \qquad\qquad (2.2.11)$$

(here we use (2.2.2)),

$$\alpha_{jj}(\tilde{x}) = \tilde{x}^{\mu^j}. \qquad\qquad (2.2.12)$$

Now

$$y_j(\tilde{x}) = \sum_{k=1}^{j} \alpha_{kj}(\tilde{x}) z_k(x) \qquad\qquad (2.2.13)$$

(cf. (2.2.3)) and $\varphi(y_j) = \varphi^j$. All the z_k are holomorphic in U^* and have at most a pole at 0 (so we may speak about $\varphi(z_k)$). Thus $z_k = x^{\varphi(z_k)} w_k(x)$ with some w_k holomorphic in U and such that $w_k(0) \neq 0$. The case $z_k \equiv 0$ is excluded here because the z_k are columns of the matrix Z which is nondegenerate, since Y is nondegenerate. It follows from (2.2.11) that

$$\varphi(\alpha_{kj} z_k) \geq \varphi(z_k). \qquad\qquad (2.2.14)$$

Indeed, if $\lambda < \varphi(z_k)$, then $\lambda + \varepsilon < \varphi(z_k)$ for some $\varepsilon > 0$, and

$$\frac{\alpha_{kj} z_k}{|x|^\lambda} = (|x|^\varepsilon \alpha_{kj}) \left(\frac{z_k}{|x|^{\lambda+\varepsilon}} \right),$$

where both factors tend to 0 as $x \to 0$ (with the usual provisos). We see that

$$\left\{ \lambda; \frac{\alpha_{kj} z_k}{|x|^\lambda} \to 0 \text{ as } x \to 0 \right\} \supset \{\lambda; \lambda < \varphi(z_k)\},$$

which implies $\varphi(\alpha_{kj} z_k) \geq \varphi(z_k)$ (cf. (2.2.4)).

Moreover,

$$\varphi(\alpha_{jj} z_j) = \varphi(z_j). \qquad\qquad (2.2.15)$$

In view of (2.2.14), we need only prove that

$$\varphi(\alpha_{jj} z_j) \leq \varphi(z_j).$$

We have

$$\frac{|\alpha_{jj} z_j|}{|z|^{\varphi(z_j)+1-\varepsilon}} = \left| \tilde{x}^{\mu^j - 1 + \varepsilon} \right| \left(\frac{|z_j|}{|x|^{\varphi(z_j)}} \right).$$

If $\varepsilon > 0$ is sufficiently small, then $\rho^j - 1 + \varepsilon < 0$, and the first factor $\left| \tilde{x}^{\mu^j - 1 + \varepsilon} \right| \to \infty$, while the second tends to $|w_j(0)| \neq 0$, as $x \to 0$. Thus

$$\sup \left\{ \lambda; \frac{\alpha_{jj} z_j}{|x|^\lambda} \to 0, \text{ as } x \to 0 \right\} \leq \varphi(z_j) + 1 - \varepsilon,$$

and $\varphi(\alpha_{jj} z_j)$, i.e., the integer part of this sup, is $\leq \varphi(z_j)$.

We claim that

$$\varphi(z_j) \geq \varphi(y_j) = \varphi^j. \tag{2.2.16}$$

As $y_1 = \alpha_{11} z_1$ (cf. (2.2.13)), this is already proved for $j = 1$ (cf. (2.2.15)); we even have

$$\varphi(z_1) = \varphi(y_1) = \varphi^1. \tag{2.2.17}$$

We proceed by induction. Assume that

$$\varphi^1 = \varphi(y_1) \leq \varphi(z_1), \ldots, \varphi^{j-1} = \varphi(y_{j-1}) \leq \varphi(z_{j-1}). \tag{2.2.18}$$

Rewrite (2.2.13) as

$$y_j = \alpha_{jj} z_j + \sum_{k=1}^{j-1} \alpha_{kj} z_k. \tag{2.2.19}$$

Assume, for contradiction, that $\varphi^j > \varphi(z_j)$. Then for the second summand (Σ) in (2.2.19) we have

$$\varphi\left(\sum_{k=1}^{j-1}\alpha_{kj}z_k\right) \geq \min\{\varphi(\alpha_{kj}z_j); k=1,\ldots,j-1\} \geq$$

$$\geq \min\{\varphi(z_k); k=1,\ldots,j-1\} \geq \min\{\varphi^k; k=1,\ldots,j-1\} =$$

$$= \varphi^{j-1} \geq \varphi^j > \varphi(z_j) = \varphi(\alpha_{jj}z_j).$$

(We use (2.2.8), (2.2.14), (2.2.18), (2.2.10), our assumption and (2.2.15)). Now in (2.2.19) we have two summands ($\alpha_{jj}z_j$ and Σ) and

$$\varphi(\alpha_{jj}z_j) < \varphi(\Sigma).$$

According to (2.2.8)

$$\varphi(y_j) = \varphi(\alpha_{jj}z_j) = \varphi(z_j)$$

(we use (2.2.15) again), which contradicts to our assumption.

As z_j is holomorphic and $\neq 0$ on U^* and as it has at most a pole at 0, (2.2.16) allows us to write

$$z_j(x) = x^{\varphi^j} v_j(x), \tag{2.2.20}$$

where $v_j(x)$ is holomorphic on U. Thus

$$Z(x) = (z_1(x),\ldots,z_p(x)) = (v_1(x),\ldots,v_p(x))x^\Phi = V(x)x^\Phi,$$

where $V = (v_1,\ldots,v_p)$ is holomorphic in U, and Φ is the diagonal matrix $\text{diag}(\varphi^1,\ldots,\varphi^p)$. This is the improvement of (2.2.3) for Levelt's solution to (1.2.4):

$$Y(\tilde{x}) = V(x)x^\Phi \tilde{x}^E. \tag{2.2.21}$$

Here V is holomorphic in U, $\text{diag}(\varphi^i)$ and E is upper-triangular. The following rephrasement of (2.2.21) is also useful:

$$(y_1,\ldots,y_p) = (v_1,\ldots,v_p)x^\Phi \tilde{x}^E. \tag{2.2.22}$$

Because of the diagonal form of Φ (hence x^Φ) and the upper triangular form of E (hence x^E), (2.2.22) can be "truncated" at any "coordinate": for any $l \in \{1,\ldots,p\}$

$$(y_1, \ldots, y_l) = (v_1, \ldots, v_l) x^{\Phi'} \tilde{x}^{E'}. \tag{2.2.23}$$

Here $\Phi' = (\mathrm{diag}\ \varphi^j)_{j=1,\ldots,l}$ and E' is the upper left $l \times l$ block of E, that is, if

$$E = (e_{ij})_{i,j=1,\ldots,p}, \text{ then } E' = (e_{ij})_{i,j=1,\ldots,l}.$$

Levelt in his paper used another basis. Here is the description of it. Let us consider a decomposition

$$X = X_1 \oplus \cdots \oplus X_s$$

of the space X into the direct sum of root subspaces X_j corresponding to different eigenvalues λ^j of G. Let us choose Levelt's basis in each X_j. A basis of X obtained by joining of these bases is called *a strongly Levelt's basis*.

Any strongly Levelt's basis $(y_1, \ldots, y_p)$ takes all values ψ^j with their multiplicities κ_j.

Indeed, otherwise there were a linear combination $w = \sum_{i=1}^{p} c_i y_i$ with the following property:

$$\varphi(w) > \min_i \varphi(c_i y_i).$$

Add all terms belonging to the same root space in the above sum and rewrite it as follows:

$$w = w_1 + \cdots + w_s,$$

where $w_i \in X_{s_i}$, $w_i \neq 0$. Note that $\min \varphi(w) = \min_i \varphi(c_i y_i) < \varphi(w)$ (this follows from the fact that $\varphi(\sum p_i y_i) = \min_i(p_i y_i)$ for any Levelt's basis $\{y_i\}$ of X_j; and the last equality, in turn, follows from the construction of a Levelt's basis).

Let s_i be the number such that

$$(\sigma^* - \lambda^i \mathrm{id})^{s_i} X_i = 0$$

and let $\varphi(w_j) = \min_i \varphi(w_i)$. Then $\varphi(w_j) < \varphi(w)$. Denote by $P(\sigma^*)$ the polynomial

$$P(\sigma^*) = \prod_{i \neq j, i=1,\ldots,s} (\sigma^* - \lambda^i \mathrm{id})^{s_i}.$$

Then

$$P(\sigma^*)w = P(\sigma^*)w_j \neq 0.$$

But according to (2.2.6) we have

$$\varphi(w) \leq \varphi(P(\sigma^*)w) = \varphi(P(\sigma^*)w_j) = \varphi(w_j)$$

(in the latter equality we use also the invertibility of the restriction of $P(\sigma^*)$ on X_j), which contradicts the inequality $\varphi(w_j) < \varphi(w)$.

So we have proved that *any strongly Levelt's basis takes all values ψ^i with their multiplicities.* As a result we obtain the following statement.

Lemma 2.2.1 *Any Levelt's basis can be obtained from some strongly Levelt's basis with help of some shuffle of its parts containing in the corresponding root subspaces and consequent upper-triangular transformation.*

If (1.2.1) has an isolated singular point at a_i and we consider the system in the neighborhood U_i of a_i, denote $U_i^* = U_i \setminus a_i$ and use the universal covering $\tilde{U}_i^* \to U_i^*$. Introducing for a moment a new independent variable $x - a_i$, we can translate to a neighborhood U of 0 as above. However, one must pay some attention to the translation of $\tilde{U}_i^*$: we cannot write something like $\tilde{x} - a_i$ without special explanation, as $\tilde{x}$ is not a number. It would be more convenient to move in another direction: from 0 to a_i. The composition

$$\tilde{U}^* \to U^* \to U_i^* \quad \tilde{x} \mapsto x \mapsto x + a_i \tag{2.2.24}$$

makes $\tilde{U}^*$ a universal covering surface for U_i^*. This does not establish an isomorphism $\tilde{U}^* \leftrightarrow \tilde{U}_i^*$ in a unique way, since it can always be changed via a deck transformation. But we may decide that from now on $\tilde{U}_i^*$ is just $\tilde{U}^*$ considered as a covering surface for U_i^* according to (2.2.24). Thus $\tilde{x}$ and $\tilde{x} + a_i$ ($\tilde{x} \in \tilde{U}^*, \tilde{x} + a_i \in \tilde{U}_i^*$) denote essentially the same "abstract" point, but considered "over $x \in U^*$" or "over $x + a_i \in U_i^*$". Similarly, $\tilde{x}$ and $\tilde{x} - a_i$ with $\tilde{x} \in \tilde{U}_i^*, \tilde{x} - a_i \in \tilde{U}^*$ denote essentially the same "abstract" point, but considered "over $x \in U_i^*$" or "over $x - a_i \in U^*$". This clarifies the meaning of such symbols as $\ln(\tilde{x} - a_i)$ and $(\tilde{x} - a_i)^{E_i}$.

Now we can write

$$Y(\tilde{x}) = V_i(x)(x - a_i)^{\Phi_i}(\tilde{x} - a_i)^{E_i} \quad (\tilde{x} \in \tilde{U}_i^*, \ x = p\tilde{x} \in U_i^*). \tag{2.2.25}$$

Here V_i is holomorphic in U_i, whereas Φ_i and E_i play the same role for the singular point a_i that Φ and E played for 0.

Return to the singularity at 0. The numbers $\beta^k := \varphi^k + \mu^k$ (where, as before, $\rho^k = Re\,\mu^k$, μ^k are the eigenvalues of E and $0 \le \rho^k < 1$) are called *exponents* of the regular system (1.2.1) at the singular point 0. One may inquire whether they are uniquely determined, as there is generally some freedom in the choice of the order in which the μ^k appear. If

$$\varphi^{k_1+\dots+k_{j-1}+1} = \dots = \varphi^{k_1+\dots+k_{j-1}+k_j} = \psi^j,$$

the numbers $2\pi i \mu^k$, where

$$k_1 + \dots + k_{j-1} + 1 \le k \le k_1 + \dots + k_{j-1} + k_j, \qquad (2.2.26)$$

are just the logarithms of the eigenvalues of σ_i^* in X^j/X^{j-1}. So their "collection" (with multiplicities) is uniquely determined, but the order in which they appear can be changed. However, to all such μ^k we add one and the same ψ^j. Thus the "collection" of β^k is uniquely determined, but there is generally some freedom in the choice of ordering. One can provide an ordering such that the sequence $Re\,\beta^k$ never increases. For groups of k corresponding to different j as in (2.2.26) this ordering is chosen independently. (If $j < l$, k satisfies (2.2.26) and m satisfies (2.2.26) with j replaced by l, then $\varphi^k = \psi^j \ge \varphi^m + 1 = \psi^l + 1$, so $Re\,\beta^k > Re\,\beta^m$ independently of the values of ρ^k and ρ^m). Let $\mu_{l_1}, \dots, \mu_{l_s}$ be the distinct eigenvalues of σ_j^* having multiplicities $n_1^j, \dots, n_s^j$. Then the basis $\bar{y}_k$, k as in (2.2.26), in which σ_j^* has an upper triangular matrix representation, can be chosen in such a way that first we take n_1^j generalized eigenvectors corresponding to μ_{l_1}, then n_2^j generalized eigenvectors corresponding to μ_{l_2}, etc. So if we order μ_{l_q} in such a way that the sequence ρ^{l_q} never increases, we achieve that $Re\,\beta^k \ge Re\,\beta^{k+1}$. After this is done (which means some additional restrictions to the Levelt's basis), the upper triangular form of E and the diagonal form of Φ in (2.2.22) imply that $Re\,\beta^k$ is just the sup λ in (2.2.4) for $y = y_k$. In this sense these numbers provide a more exact characterization of the growth of the y_k than the φ^k do. Using the $Re\,\beta^k$ we neglect only polynomials in $\ln \tilde{x}$, while using the φ^k we neglect also fractional powers of $\tilde{x}$. However, we shall not use the β^k in this role; we shall use them in a different way.

We shall often deal with the matrix

$$L(x) := \Phi + x^{\Phi} E x^{-\Phi}. \qquad (2.2.27)$$

Clearly, it is holomorphic in U^*; let us check that it is holomorphic in all of U, i.e., that the limit $L(0) := \lim_{x \to 0} L(x)$ exists. Essentially this means the existence of $\lim x^\Phi E x^{-\Phi}$. Clearly the (i, j)th coefficient of the latter is $x^{\varphi_i} e_{ij} x^{-\varphi_j}$, where $(e_{ij}) = E$. Now $e_{ij} = 0$, when $i > j$, and $\varphi_i \geq \varphi_j$, when $i \leq j$; hence

$$\lim x^{\varphi_i} e_{ij} x^{-\varphi_j} = \left\{ \begin{array}{llll} 0, & \text{when} & i < j; \\ e_{ij}, & \text{when} & i \leq j; \text{ and} & \varphi_i = \varphi_j; \\ 0, & \text{when} & i < j; \text{ and} & \varphi_i > \varphi_j. \end{array} \right.$$

Not only we have proven the existence of the limit, but we also have some information about the structure of L: *this matrix is obtained by picking the diagonal blocks out of the matrix $\Phi + E$.* Here the block structure of $\Phi + E$ is just the structure corresponding to the filtration (2.2.9) and the construction of the Levelt's basis. Using the previous notation, the (r, s) block is a $k_r \times k_s$ matrix. The diagonal elements of $L(0)$ are β^j and the trace

$$tr L(0) = tr \Phi + tr E. \qquad (2.2.28)$$

Theorem 2.2.1 (Levelt, [Le]) *The regular system (1.2.1) is Fuchsian (at 0) if and only if $V(0)$ (cf.(2.2.21)) is invertible.*[2]

Proof. Let us begin at "if" part (this is easier and, by the way, provides a useful formula). (1.2.4) implies that $A(x) = \dot{Y} Y^{-1}$ (where dot denotes differentiation). Take Levelt's Y and use (2.2.21), bearing in mind that $(x^\Phi)^{\cdot} = \frac{1}{x} \Phi x^\Phi$ and analogously for $\tilde{x}^E$. We get

$$\dot{Y} = \dot{V} x^\Phi \tilde{x}^E + \frac{1}{x} V \Phi x^\Phi \tilde{x}^E + \frac{1}{x} V x^\Phi E \tilde{x}^E = \frac{1}{x} (x \dot{V} + V L) x^\Phi \tilde{x}^E.$$

But $Y^{-1} = (\tilde{x}^E)^{-1} (x^\Phi)^{-1} V^{-1}$, so

---

[2]F.R.Gantmacher in his well-known book [Ga] has a theorem which contains the essential part of the theorem 1 (chapter XIV, §10, Theorem 2. We warn the reader of this book about a difference in terminology: Gantmacher calls a system regular at point a if, in our terms, it is Fuchsian there. Gantmacher has no special term for the system which we call regular). His theorem asserts that if the system (1.2.1) is Fuchsian at 0 then (1.2.4) has a solution of the form (2.2.21) with $V(0) =$ identity matrix and Φ is diagonalizable with integer eigenvalues (while E is "responsible" for the monodromy). However, he does not use the Levelt's valuation φ and does not characterize Levelt's fundamental system of solutions in terms of this valuation.

$$A = \dot{Y}Y^{-1} = \frac{1}{x}(x\dot{V} + VL)V^{-1}. \tag{2.2.29}$$

All matrices on the right hand side are holomorphic in U (here we use the invertibility of $V(0)$)). So A indeed has (no more than) a pole of first order at 0.

We turn to the "only if" part. We already know that $v_1(0) \neq 0$; cf. (2.2.17) and (2.2.20) (Having in mind also that z_1 and v_1 are single-valued, thus representable by some Laurent and Taylor series). This is true even for regular systems. Now let (1.2.1) be Fuchsian, i.e.,

$$A(x) = \frac{1}{x}C(x), \tag{2.2.30}$$

where C is holomorphic in U. In view of (2.2.29), $CV = x\dot{V} + VL$. Passing to the limit as $x \to 0$ yields

$$C(0)V(0) = V(0)L(0). \tag{2.2.31}$$

This implies $L(0)\mathrm{Ker}V(0) \subset \mathrm{Ker}V(0)$. Indeed, if $V(0)z = 0$, then $V(0)L(0)z = C(0)V(0)z = 0$. It follows that

$$\tilde{x}^{L(0)}\mathrm{Ker}V(0) \subset \mathrm{Ker}V(0). \tag{2.2.32}$$

Now assume that $c \in \mathrm{Ker}V(0)$, $c \neq 0$. Consider the solution $y = Y(\tilde{x})c$ to (1.2.1). Let $\varphi(y) = \psi^m$. We shall prove that $\varphi(y) > \psi^m$, which is a contradiction. As

$$y = Y(\tilde{x})c = V(x)x^{\Phi}\tilde{x}^E c =$$
$$= V(x)\tilde{x}^{L(0)}c + V(x)\left(x^{\Phi}\tilde{x}^E - \tilde{x}^{L(0)}\right)c,$$

it is sufficient to prove that

$$\varphi\left(V(x)\tilde{x}^{L(0)}c\right) > \psi^m, \tag{2.2.33}$$

$$\varphi\left(V(x)\left(x^{\Phi}\tilde{x}^E - \tilde{x}^{L(0)}\right)c\right) > \psi^m. \tag{2.2.34}$$

It follows from $c \in \mathrm{Ker}\,V(0)$ and (2.2.31) that $V(0)\tilde{x}^{L(0)}c = 0$. Now it is clear that

$$V(x)\tilde{x}^{L(0)}c = (V(0) + O(x))\tilde{x}^{L(0)}c = O(x)\tilde{x}^{L(0)}c. \tag{2.2.35}$$

Repeat once more that $L(0)$ is obtained form $\Phi + E$ by picking only diagonal blocks. It is assumed that we use the block structure corresponding to the filtration (2.2.9) and the choice of the Levelt's basis. The "size" of the i-th block is k_i, and I_i is the identity matrix of order k_i, the i-th (diagonal) block of $L(0)$ is $\psi^i I_i + E_{ii}$, where E_{ii} is the corresponding block of E. Thus $\tilde{x}^{L(0)}$ consists of the diagonal blocks

$$\tilde{x}^{\psi^i I_i + E_{ii}} = \tilde{x}^{\psi^i}\tilde{x}^{E_{ii}}.$$

Let $c = \begin{pmatrix} c_1 \\ \vdots \\ c_m \\ 0 \\ \vdots \\ 0 \end{pmatrix}$, $c_m \neq 0$, be the corresponding representation of c. Then

$$\tilde{x}^{L(0)}c = \left(x^{\psi^1}\tilde{x}^{E_{11}}c_1, \ldots, x^{\psi^m}\tilde{x}^{E_{mm}}c_m, 0, \ldots, 0\right).$$

But for the coefficients of $\tilde{x}^E$ (2.2.11) holds. So $\varphi\left(\tilde{x}^{L(0)}c\right) \geq \psi^m$. Now using (2.2.35) we obtain (2.2.33).

We turn to (2.2.34). Since $V(x)$ is holomorphic at $x = c$,

$$\varphi\left(V(x)\left(x^\Phi\tilde{x}^E - \tilde{x}^{L(0)}\right)c\right) \geq \varphi\left(\left(x^\Phi\tilde{x}^E - \tilde{x}^{L(0)}\right)c\right),$$

and it is sufficient to prove that

$$\varphi\left(\left(x^\Phi\tilde{x}^E - \tilde{x}^{L(0)}\right)c\right) > \psi^m. \tag{2.2.36}$$

$x^\Phi\tilde{x}^E$ has the following block structure

$$x^\Phi\tilde{x}^E = \begin{pmatrix} x^{\psi^1}\left(\tilde{x}^E\right)_{11} & x^{\psi^1}\left(\tilde{x}^E\right)_{12} & \cdots & x^{\psi^1}\left(\tilde{x}^E\right)_{1h} \\ 0 & x^{\psi^2}\left(\tilde{x}^E\right)_{22} & \cdots & x^{\psi^2}\left(\tilde{x}^E\right)_{2h} \\ \cdots & \cdots & \cdots & \cdots \\ 0 & 0 & \cdots & x^{\psi^h}\left(\tilde{x}^E\right)_{hh} \end{pmatrix}. \tag{2.2.37}$$

Clearly $(\tilde{x}^E)_{ii} = \tilde{x}^{E_{ii}}$, as E is upper-triangular. Indeed, if A and B are of such type then $(AB)_{ii} = A_{ii}B_{ii}$. It follows that for any polynomial p

$$p(E)_{ii} = p(E_{ii}),$$

and then the same is true for any "function of matrix" $f(E)$. (Essentially it is not the upper triangular form of E which is used here, but only the fact that all blocks below the diagonal blocks are zeroes: $E_{ij} = 0$ for $i > j$). So the diagonal blocks in (2.2.37) are the same as in $\tilde{x}^{L(0)}$, and $x^\Phi \tilde{x}^E - \tilde{x}^{L(0)}$ consists of the following blocks:

$$\left(x^\Phi \tilde{x}^E - \tilde{x}^{L(0)}\right)_{ij} = 0 \text{ for } i \geq j,$$
$$\left(x^\Phi \tilde{x}^E - \tilde{x}^{L(0)}\right)_{ij} = x^{\psi^i}\left(\tilde{x}^E\right)_{ij} \text{ for } i < j.$$

When we act by this matrix on c, we get a column vector $z = (z_1, \ldots, z_{m-1}, 0, \ldots, 0)$ with

$$z_i = \sum_{j=i+1}^{m} x^{\psi^j}\left(\tilde{x}^E\right)_{ij} c_j.$$

If $m = 1$, we get $z = 0$, $\varphi(z) = \infty > \psi^m$. If $m > 1$, here figure only $x^{\psi^1}, \ldots, x^{\psi^{m-1}}$. Again referring to (2.2.11), we conclude that $\varphi(z) \geq \psi^{m-1} > \psi^m$. In any case we arrive at (2.2.34). The theorem is proved.

The following useful statement is the direct corollary of formula (2.2.29).

Corollary 2.2.1 *Let a solution $Y(\tilde{x})$ (not necessary a Levelt's one) to (1.2.4) have a factorization of form (2.2.21) with some matrix Φ with integer coefficients. Let $V(x)$ be holomorphically invertible at $x = 0$ and let $L(x)$ from (2.2.27) be holomorphic. Then system (1.2.4) is Fuchsian at $x = 0$.*

At first glance, Levelt's theory may seem to be "nonconstructive" – at any rate, less constructive than a more computational classical approach, where we substitute series into the system and try to extract useful information from the relations thus obtained. Nonetheless, Levelt's theory provides some "explicit" information, and rather quickly. Consider the Fuchsian system (1.2.1), (2.2.30). Since $V(0)$ is invertible, it follows from (2.2.31) that the matrices $C(0)$ and $L(0)$ are similar. Thus the numbers β^j are just the eigenvalues of $C(0)$. Knowing them, one can find $Re(\varphi^j + \mu^j) = \varphi^j + \rho^j$, $\varphi^j = [Re\,\beta^j]$ (integer part), ρ^j and μ^j. Also, $L(0)$

has the same Jordan normal form as $C(0)$ (although in a different basis which remains unknown). It is easy to see that then the same is valid for $\Phi + E$ (but the corresponding basis may be different from the previous two). However, this need not define the Jordan normal form of E as the following example shows. (We shall meet this example again in Section 2.3). Let $p = 2$, $\varphi^1 = 1$, $\varphi^2 = -1$, $E = \begin{pmatrix} 0 & e \\ 0 & 0 \end{pmatrix}$.

Then $\Phi + E = \begin{pmatrix} 1 & e \\ 0 & -1 \end{pmatrix}$, and since $\varphi^1 \neq \varphi^2$, $L(0) = \begin{pmatrix} 1 & 0 \\ 0 & -1 \end{pmatrix}$. Now

assume that we are given a system (1.2.1), (2.2.30) with $C(0) = \begin{pmatrix} 1 & 0 \\ 0 & -1 \end{pmatrix}$. It

follows that $L(0)$ has the same form in some basis, but it remains uncertain whether $e = 0$ or $e \neq 0$ – any version leads to our $L(0)$.

Let us also mention that (cf. (2.2.28))

$$\sum_{j=1}^{p} \beta^j = tr(\Phi + E) = tr\, L(0) = tr\, C(0). \tag{2.2.38}$$

Another important theorem by Levelt is of a more global character. Consider a system (1.2.1) which is regular on $\bar{\mathbb{C}}$ with singularities $a_1, \ldots, a_n$. Near any a_i we apply the previous theory (using a local parameter $x - a_i$ or $1/x$ if $a_i = \infty$) and obtain the corresponding matrices and numbers. Now they depend on i, so we write

$$\Phi_i, E_i, L_i(0), \mu_i^j, \rho_i^j, \varphi_i^j, \beta_i^j. \tag{2.2.39}$$

It is worth mentioning that at the moment no global considerations are needed yet – near a_i one works in a small circular neighborhood U_i. (If $a_i = \infty$, U_i is properly a circular neighborhood in terms of the variable $t = 1/x$; in terms of x it is the complement to a large disk containing all other singularities). The only global remark at the moment is that in every U_i we use the orientation induced by the standard orientation of $\bar{\mathbb{C}}$, and the deck transformation σ_i of the universal covering $\tilde{U}_i^* \rightarrow U_i^* = U^i \setminus a_i$ corresponds to one turn around a_i in a positive direction. (That is, counterclockwise for finite a_i. With regard to the singular point at ∞, if there is such a singular point, the positive direction of a trip around it is to be understood in a sense that is positive (counterclockwise) in terms of the local parameter $t = 1/x$. In terms of the initial variable x, this means a turn along a sufficiently large circle (surrounding all other singularities) in the negative direction, i.e., clockwise).

Theorem 2.2.2 (Levelt, [Le]) *For any regular system $\sum_{i,j} \beta_i^j \leq 0$. Equality holds if and only if the system is Fuchsian.*

The proof of Theorem 2.2.2 is easy. We may assume that the singularities $a_1, \ldots, a_n$ are different from ∞ (this can be achieved always by a suitable change of the independent variable).

Let $\tilde{Y}$ be a fundamental matrix to (1.2.1), then for the matrix differential form $w = A(x)dx$ we have

$$tr\, w = d \ln \det Y. \qquad (2.2.40)$$

Since for arbitrary i : $\tilde{Y}(\tilde{x}) = Y(\tilde{x})S_i$, where $\det S_i \neq 0$ and $Y(\tilde{x})$ is a Levelt's fundamental matrix in a neighborhood U_i^*, we obtain from (2.2.25) and (2.2.38):

$$\operatorname{res}_{a_i} d \ln \det \tilde{Y} = \sum_{j=1}^{p} \beta_i^j + b_i, \qquad (2.2.41)$$

where $b_i = \det V_i(a_i) \geq 0$.

By the theorem on the sum of residues, applied to the form $tr\, w$ from (2.2.40)

$$\sum_{i,j} \beta_i^j + \sum_{i=1}^{n} b_i = 0,$$

thus $\sum_{i,j} \beta_i^j \leq 0$ and $\sum_{i,j} \beta_i^j = 0$ if and only if $b_1 = \cdots = b_n = 0$. But the latter equalities imply $\det V_i(a_i) \neq 0$ for all a_i. Theorem 2.2.2 follows now from Theorem 2.2.1.

In this argument there was no need to enter into the relationship between the local description of the solutions provided by Levelt's theory and their global behavior. However, this will be necessary further on, and we shall finish this section by discussing this subject, as the coordination of the local and global points of view at all.

In the local theory one uses such functions as $\ln(x - a_i)$ or $(x - a_i)^B = \exp(B \ln(x - a_i))$, where B is some matrix. These are multivalued functions of x, and one must consider them on the appropriate Riemann surface. In the local theory we remain near a_i and so we use only a part of this surface. All we need is the universal covering $p_i : \tilde{U}_i^* \to U_i$ for U_i^* which are as above. Solution of (1.2.1) and (1.2.4) are correctly defined there as well as the functions mentioned above.

In the global theory things are different – we consider our solution on $\tilde{S}$, while $\ln(x - a_i)$ has another Riemann surface $\tilde{T}_i$ covering $\mathbb{C} \setminus \{a_i\}$. Of course we shall use $\ln(x - a_i)$ only when we analyze the behavior of our solutions near a_i. However, this does not make things completely local in the following respect: we must remember that many "branches" of our solutions near a_i are obtained via continuations along paths which are not completely contained in U_i^* and are not homotopic to the path from U_i^*.

So let us consider this situation in more detail. Now S and $p : \tilde{S} \to S$ will be as in the very beginning of Section 2.2, while U_i, U_i^* and $p_i : \tilde{U}_i^* \to U_i^*$ will play the same role for a_i, that U, U^* and $\tilde{U}^* \to U^*$ play for the singular point 0. If $n > 2$, then $p^{-1}U_i^*$ is much larger than $\tilde{U}_i^*$. (This corresponds to the fact that $Y(\tilde{x})$ can be analytically continued not only along paths lying in U_i^*, but also along paths going around other singular points a_j, and this may provide new "elements" of the multivalued function Y). Namely, $p^{-1}U_i^*$ is disconnected and each of its connected components is isomorphic to $\tilde{U}_i^*$ as a covering space over U_i^*. So we can identify one of this components with $\tilde{U}_i^*$, but, of course, after we have done this we must distinguish between $\tilde{U}_i^*$ and the other components. In order to fix somehow this identification (or just to describe it in a more concrete way), let us choose points $b \in S$, $u_i \in U_i^*$ and some paths $\beta_1, \ldots, \beta_n$ connecting b to $u_1, \ldots, u_n$. Each $\tilde{x} \in \tilde{U}_i^*$ can be interpreted as a class of mutually homotopic paths $\{\gamma\}$ in U_i beginning at u_i and ending at $x = p_i \tilde{x}$ (here the homotopy is the homotopy with the fixed ends). Analogously, each point of $\tilde{S}$ also can be interpreted as some class $\{\delta\}$ of mutually homotopic paths on S beginning at b and all having the same end point. (It is precisely this interpretation that allows one to define a left action of $\pi_1(S, b)$ on $\tilde{S}$ and thus establish the isomorphism

$$\pi_1(S, b) \cong \Delta. \tag{2.2.42}$$

For $\{\delta\}$ as before and $\varepsilon \in \pi_1(S, b)$ define $\{\varepsilon\}\{\delta\} = \{\varepsilon\delta\}$). Then we can define a map $\tilde{U}_i^* \hookrightarrow \tilde{S}$ by $\{\gamma\} \mapsto \{\beta_i\gamma\}$. Any other connected component of $p^{-1}(U_i^*)$ can be obtained as $\tau\tilde{U}_i^*$ with some $\tau \in \Delta$ (which is by no means unique, see below).

Let α_i be a loop (a closed path) in U_i^* beginning and ending at u_i and making one turn around a_i in the positive direction. It defines a deck transformation σ_i: $\tilde{U}_i^* \to \tilde{U}_i^*$ (which we have already used): $\sigma_i\{\gamma\} = \{\alpha_i\gamma\}$. This formula does not define a transformation of $\tilde{S}$, because instead of γ we must use paths δ in S beginning at b while α_i ends at a different point u_i. The closest meaningful analog is $\{\delta\} \mapsto \{\beta_i\alpha_i\beta_i^{-1}\delta\}$. For $\{\delta\} = \{\beta_i\gamma\}$ it gives $\{\delta\} \mapsto \{\beta_i\alpha_i\gamma\}$, so this is really an extension of σ_i to all of $\tilde{S}$. In this way we can consider σ_i (and other deck transformations σ_i^k of $\tilde{U}_i^*$) as deck transformations of the whole $\tilde{S}$. Solutions y to

(1.2.1) or Y to (1.2.4) which where considered in the local theory "near a_i" (thus defined only on $\tilde{U}_i^*$) can be extended (analytically) over the whole $\tilde{S}$, thus becoming a global solution. Conversely, any global solution y or Y (defined on $\tilde{S}$) can be restricted to $\tilde{U}_i^*$ and this $y \mid \tilde{U}_i^*$ or $Y \mid \tilde{U}_i^*$ is an object of the local theory. However, this does not mean the same as to say: "y or Y considered as a multivalued function of x for $x \in U_i^*$" because (generally) y, Y have near a_i other branches as well. With this precaution in mind we can say that there exists an isomorphism

$$X_i \longleftrightarrow X, \qquad\qquad (2.2.43)$$

where the elements of X are local solutions to (1.2.1) defined on $\tilde{U}_i^*$ (previously we considered only the singularity and denoted this space by X), while the elements of X are global solutions to (1.2.1) defined on $\tilde{S}$. (This isomorphism depends upon our identification of $\tilde{U}_i^*$ with a component of $p^{-1}U_i^*$, i.e. on β_i). Filtration and valuation in X_i depends on i, so we write $\varphi_i(y)$, X_i^j (cf. the use of i in (2.2.39)). The isomorphism (2.2.43) allows one to consider φ_i as a valuation of X and X_i^j as filtration of X (but there they somewhat depend on our choice of the identification of $\tilde{U}_i^*$ with a component of $p^{-1}(U_i^*)$, i.e. on β_i. It may well happen that after continuing a local solution y (initially defined on $\tilde{U}_i^*$) along a path going around another a_j one obtains a new local solution having another order of growth as $x \to a_i$. Note that φ_i is invariant under σ_i, but it need not be invariant with respect to all of Δ). Levelt's basis of X_i gives us (due to (2.2.43)) a fundamental system Y of global solutions, but generally only $Y \mid \tilde{U}_i^*$ is a Levelt's basis[3] and has a representation (2.2.25). For the point $\tilde{x}$ of another connected component $\tau \tilde{U}_i^*$ of $p^{-1}(U_i^*)$ we have

$$Y(\tilde{x}) = Y\left(\tau^{-1}\tilde{x}\right)\chi\left(\tau^{-1}\right) = V(x)(x - a_i)^{\Phi_i}\left(\tau^{-1}\tilde{x} - a_i\right)^{E_i}\chi\left(\tau^{-1}\right). \quad (2.2.44)$$

Clearly τ and $\tau\sigma_i^k$, $k \in \mathbb{Z}$, define the same connected component of $p^{-1}(U_i^*)$: $\tau\tilde{U}_i^* = \tau\sigma_i^k\tilde{U}_i^*$, since $\sigma_i\tilde{U}_i^* = \tilde{U}_i^*$. Thus there is some ambiguity when we speak about τ defining a certain component. (The only ambiguity which is possible here is the ambiguity with τ and $\tau\sigma_i^k$: if $\tau\tilde{U}_i^* = \tau_1\tilde{U}_i^*$, then $\tau_1 = \tau\sigma_i^k$ with some integer k. See Chapter 3). Hence there is also some ambiguity in the factors $\left(\tau^{-1}\tilde{x} - a_i\right)^{E_i}$, $\chi\left(\tau^{-1}\right)$. However, there is no ambiguity in the product: if $\tilde{x} = \tau\tilde{x}_0 = \tau_1\tilde{x}_1$, $\tilde{x}_0 \in \tilde{U}_i^*$, $\tilde{x}_1 \in \tilde{U}_i^*$, then

---

[3]Formally, the notion of being a Levelt's basis is defined only for a system of p vector functions defined on $\tilde{U}_i^*$. However, if $\tilde{x} \in p^{-1}(U_i^*) \setminus \tilde{U}_i^*$, we can take a germ of Y at $\tilde{x}$ and consider this germ as a germ of a matrix-valued function at $x = p\tilde{x}$. This function admits an analytic continuation in U_i^*, and the resulting complete multivalued function can be considered as a function Y_1 on $\tilde{U}_i^*$. If $Y \mid \tilde{U}_i^*$ is a Levelt's basis, even though Y_1 need not be such a basis.

$$(\tilde{x}_0 - a_i)^{E_i} \chi \left(\tau^{-1}\right) = (\tilde{x}_1 - a_i)^{E_i} \chi \left(\tau_1^{-1}\right).$$

We postpone the proof of this fact till Chapter 3, where it is essential. Now temporary we shall be more careless. We simply select for any connected component of $p^{-1}U_i^*$ a τ such that this component equals to $\tau\tilde{U}_i^*$, and define on this component

$$\ln(\tilde{x} - a_i) := \ln\left(\tau^{-1}\tilde{x} - a_i\right).$$

Then instead of (2.2.44) we can write "in a more classical style"

$$Y(\tilde{x}) = V(x)(x - a_i)^{\Phi_i}(\tilde{x}_0 - a_i)^{E_i} \chi \left(\tau^{-1}\right). \tag{2.2.45}$$

Note that on $\tilde{U}_i^*$

$$\ln(\sigma_i\tilde{x} - a_i) = \ln(\tilde{x} - a_i) + 2\pi i, \tag{2.2.46}$$

which is the well-known classical fact. But whether (2.2.46) holds for all $\tilde{x} \in p^{-1}U_i^*$ depends on the choice of our τ's. We shall not discuss this, as (2.2.46) is used in the Chapter 3 only for $\tilde{x} \in \tilde{U}_i^*$.

2.3 The second order system

We are going to study the system (2.1.6) with B as in (2.1.4). Computation shows that for this system the condition analogous to (1.2.3) is fulfilled. So the only singularities are those which, so to say, manifest themselves explicitly in (2.1.4), i.e. $a_1 = 0$, $a_2 = -1$, $a_3 = 1$, $a_4 = 1/2$. The situation with a_2, a_3, a_4 is easy. The corresponding B_i are

$$\frac{1}{6}\begin{pmatrix} -1 & 1 \\ -1 & 1 \end{pmatrix}, \frac{1}{2}\begin{pmatrix} -1 & -1 \\ 1 & 1 \end{pmatrix}, \frac{1}{3}\begin{pmatrix} -1 & 1 \\ -1 & 1 \end{pmatrix}. \tag{2.3.1}$$

Clearly they are degenerate and their traces are 0, so the characteristic equation $\lambda^2 - \lambda \, tr \, B_i + \det B_i = 0$ reduces to $\lambda^2 = 0$. Hence these matrices are nilpotent and, of course, they are of rank 1, as any nonzero degenerate (2×2)-matrix. It follows that the block structure of $L_i(0)$ is as follows: all of $L_i(0)$ is one diagonal block. This implies that $\Phi_i + E_i = L_i(0)$, indeed there are no blocks to annihilate when passing from $\Phi_i + E_i$ to $L_i(0)$. Consequently,

$$\varphi_i^j = \mu_i^j = \rho_i^j = \beta_i^j = 0, \quad i = 2, 3, 4; \ j = 1, 2. \tag{2.3.2}$$

Note also that $\Phi_i = 0$ ($i = 2, 3, 4$) and we need not write the factors x^{Φ_i} in the corresponding representation (2.2.44) or (2.2.45) near a_2, a_3, a_4. The filtration (2.2.9) reduces to the trivial one: $0 \subset X_1 = X$ and the condition $\varphi(y_i) = \varphi^j$ imposes no restriction on the choice of y_j. The only restriction which remains is that E_i in this basis has the form $\begin{pmatrix} 0 & e_i \\ 0 & 0 \end{pmatrix}$. However, for any fundamental system of solution Y near a_i we have a representation

$$Y(\tilde{x}) = V_i(x)(\tilde{x} - a_i)^{E_i}, \ \tilde{x} \in \tilde{U}_i^* \tag{2.3.3}$$

with some E_i nilpotent of rank 1, V_i holomorphic on a small circular neighborhood U_i of a_i and $\det V_i(0) \neq 0$. Indeed, let Y_i be a Levelt's fundamental system of solutions. Then $Y = Y_i C$ with some constant invertible C. This implies that

$$Y = V_i(\tilde{x} - a_i)^{E_i} C = V_i C C^{-1}(\tilde{x} - a_i)^{E_i} C = (V_i C)(\tilde{x} - a_i)^{C^{-1} E_i C}.$$

Clearly $V_i C$ and $C^{-1} E_i C$ have the properties just described. Finally, in $p^{-1}(U_i^*) \backslash \tilde{U}_i^*$

$$Y(\tilde{x}) = V_i(x) \left(\tau^{-1} \tilde{x} - a_i \right)^{E_i} \chi \left(\tau^{-1} \right), \ \text{if } \tilde{x} \in \tau \tilde{U}_i^*, \tag{2.3.4}$$

where one can replace $\tau^{-1} \tilde{x} - a_i$ by $\tilde{x} - a_i$ for appropriate τ.

Near $a_1 = 0$ the situation is a little less trivial. $L_1(0)$ is similar to $B_1 = \begin{pmatrix} 1 & 0 \\ 0 & -1 \end{pmatrix}$, so

$$\varphi_1^1 = 1, \ \varphi_1^2 = -1, \ \mu_1^j = \rho_1^j = 0 \ (j = 1, 2); \ \beta_1^1 = 1, \ \beta_1^2 = -1. \tag{2.3.5}$$

But this is exactly the example discussed in Section 2.2, so the Jordan canonical form of E_1 remains unknown. In order to find it we shall consider the second order terms in $C(x) := xB(x)$ and $V_1(x)$.

Rewrite (2.1.4) as follows:

$$B(x) = \left(\frac{1}{x} - \frac{1}{6(x+1)} - \frac{1}{2(x-1)} - \frac{1}{3(x-1/2)} \right) \Phi +$$
$$\left(\frac{1}{6(x+1)} - \frac{1}{2(x-1)} - \frac{1}{3(x-1/2)} \right) \Psi,$$

where $\Phi = \begin{pmatrix} 1 & 0 \\ 0 & -1 \end{pmatrix}$, $\Psi = \begin{pmatrix} 0 & 1 \\ -1 & 0 \end{pmatrix}$. (The notation Φ is in accordance with the notation used in Section 2.2: it is the same $\Phi = \Phi_1$ which described the valuation at $a_1 = 0$, cf. (2.3.5)). Expanding the fractions at $x = 0$, we get

$$C(x) = (1 + x + 2x^2)\Phi - x^2\Psi + \dots \tag{2.3.6}$$

Since E_1 must be upper triangular and $\mu_1^j = 0$, we write $E_1 = \begin{pmatrix} 0 & e \\ 0 & 0 \end{pmatrix}$. We shall use the following formulas which are easily verified:

$$\left[\Phi, \begin{pmatrix} a & b \\ c & d \end{pmatrix}\right] = \Phi\begin{pmatrix} a & b \\ c & d \end{pmatrix} - \begin{pmatrix} a & b \\ c & d \end{pmatrix}\Phi = \begin{pmatrix} 0 & 2b \\ -2c & 0 \end{pmatrix}, \tag{2.3.7}$$

$$\begin{pmatrix} \alpha & 0 \\ 0 & \beta \end{pmatrix} E_1 \begin{pmatrix} \alpha^{-1} & 0 \\ 0 & \beta^{-1} \end{pmatrix} = \begin{pmatrix} 0 & \alpha\beta^{-1}e \\ 0 & 0 \end{pmatrix} = \alpha\beta^{-1}E_1, \tag{2.3.8}$$

$$x^\Phi E_1 x^{-\Phi} = x^2 E_1. \tag{2.3.9}$$

(The first formula is a commutator of matrices and the last follows from (2.3.8) with $\alpha = x$, $\beta = x^{-1}$). Hence $L_1(x) = \Phi + x^2 E_1$ (cf.(2.2.27)). Since V_1 is holomorphic in U_1, write

$$V_1 = \sum_{n=0}^{\infty} x^n W_n. \tag{2.3.10}$$

Substitute these L and V, as well as C from (2.3.6), into the formula $x\dot{V} = CV - VL$ (which is equivalent to (2.2.29); in our case A has to be replaced by B and so C by (2.3.6)) and compare the coefficients of x^0, x^1, x^2. This gives

$$0 = [\Phi, W_0] \tag{2.3.11}$$

(which we already know: it is (2.2.31) with our $V_1(0) = W_0$, $C(0) = L_1(0) = \Phi$);

$$W_1 = [\Phi, W_1] + \Phi W_0; \tag{2.3.12}$$

$$2W_2 = [\Phi, W_2] + \Phi W_1 + (2\Phi - \Psi)W_0 - W_0 E_1. \qquad (2.3.13)$$

Let $W_0 = \begin{pmatrix} a & b \\ c & d \end{pmatrix}$. Then (2.3.7) and (2.3.11) yield $b = 0$, $c = 0$. As we know, there is some freedom in the choice of Levelt's basis. In any case, if (y_1, y_2) is a such a basis, then $(c_1 y_1, c_2 y_2)$ is also a Levelt's basis. (Indeed, $\varphi_1(c_i y_i) = \varphi_1(y_i)$ for $c_i \neq 0$, and if the matrix representation of σ_1^* in the first basis was upper triangular, it will remain so with respect to the second basis). The question whether $e = 0$ or $e \neq 0$ is equivalent to the question whether σ^* is the identity transformation or not; this does not depend on the basis. So replace y_1 by $\frac{1}{a}y_1$ and y_2 by $\frac{1}{d}y_2$; denote these new solutions to (2.1.6) by y_i again, that is take a new Levelt's matrix $Y \begin{pmatrix} a^{-1} & 0 \\ 0 & d^{-1} \end{pmatrix}$.

If Y satisfies (2.2.21), then

$$Y \begin{pmatrix} a^{-1} & 0 \\ 0 & d^{-1} \end{pmatrix} = V \begin{pmatrix} a^{-1} & 0 \\ 0 & d^{-1} \end{pmatrix} \begin{pmatrix} a & 0 \\ 0 & d \end{pmatrix} x^{\Phi} \begin{pmatrix} a^{-1} & 0 \\ 0 & d^{-1} \end{pmatrix} \times$$
$$\times \begin{pmatrix} a & 0 \\ 0 & d \end{pmatrix} \tilde{x}^{E_1} \begin{pmatrix} a^{-1} & 0 \\ 0 & d^{-1} \end{pmatrix}.$$

Here $\begin{pmatrix} a & 0 \\ 0 & d \end{pmatrix} x^{\Phi} \begin{pmatrix} a^{-1} & 0 \\ 0 & d^{-1} \end{pmatrix} = x^{\begin{pmatrix} a & 0 \\ 0 & d \end{pmatrix} \Phi \begin{pmatrix} a^{-1} & 0 \\ 0 & d^{-1} \end{pmatrix}} = x^{\Phi}$, bec-

ause Φ is diagonal, and $\begin{pmatrix} a & 0 \\ 0 & d \end{pmatrix} \tilde{x}^{E_1} \begin{pmatrix} a^{-1} & 0 \\ 0 & d^{-1} \end{pmatrix} = \tilde{x}^{ad^{-1}E_1}$ (cf. (2.3.8)).

Thus for the new Levelt's matrix (which we shall denote by Y again) we have (2.2.21) with V replaced by $V_1 \begin{pmatrix} a^{-1} & 0 \\ 0 & d^{-1} \end{pmatrix}$ (denoted by V_1 again), E replaced by $ad^{-1}E_1$ (denoted by $E_1 = \begin{pmatrix} 0 & e \\ 0 & 0 \end{pmatrix}$ again) and $\Phi = \Phi_1$ unchanged. But for new V_1 we still have (2.3.10) with new $W_0 = V_1(0) = I$.

Substitute this W_0 into (2.3.12). Then

$$W_1 = [\Phi_1, W_1] + \Phi. \qquad (2.3.14)$$

Let $W_1 = \begin{pmatrix} a & b \\ c & d \end{pmatrix}$. Then the elements of the second diagonal in the left hand side of (2.3.12) are (b, c), while in the right hand side $(2b, -2c)$ (cf.(2.3.7)). Thus $b = c = 0$, $[\Phi, W_1] = 0$ and (2.3.14) reduces to $W_1 = \Phi$.

Substitute these W_0 and W_1 into (2.3.13). Then

$$2W_2 = [\Phi, W_2] + \Phi^2 + 2\Phi - \Psi - E_1.$$

Consider the $(1,2)$ elements of the left and right hand sides, i.e. the elements occurring in the upper right corner there. Let $W_2 = \begin{pmatrix} \cdot & b \\ \cdot & \cdot \end{pmatrix}$. This gives the equation $2b = 2b - 1 - e$. Consequently, $e = -1 \neq 0$.

We see that E_1, like the other E_i, is a nilpotent matrix of rank 1 (and that $E_1 = \begin{pmatrix} 0 & -1 \\ 0 & 0 \end{pmatrix}$ for our Y, i.e. Levelt's Y normalized by the requirement $V_1(0) = I$). We also see that

$$V_1(x) = \begin{pmatrix} 1 + \cdots & ax^2 + \cdots \\ \beta x^2 + \cdots & 1 + \cdots \end{pmatrix}, \tag{2.3.15}$$

since the second diagonal in W_0 and W_1 is $(0,0)$.

Knowing Φ_1, E_1 and V_1, we can compute the lowest terms in the Levelt's representation for $Y(\tilde{x})$, $\tilde{x} \in \tilde{U}_1^*$ which we shall use in the next section. We have

$$e^{tE_1} = I + tE_1 + \frac{1}{2}(tE_1)^2 + \ldots = I + tE_1,$$

$$\tilde{x}^{E_1} = e^{\ln \tilde{x} \cdot E_1} = I + \ln \tilde{x} \cdot E_1 = \begin{pmatrix} 1 & -\ln \tilde{x} \\ 0 & 1 \end{pmatrix}.$$

Thus,

$$Y(\tilde{x}) = \begin{pmatrix} 1 + \cdots & \alpha x^2 + \ldots \\ \beta x^2 + \cdots & 1 + \cdots \end{pmatrix} \begin{pmatrix} x & 0 \\ 0 & x^{-1} \end{pmatrix} \begin{pmatrix} 1 & -\ln \tilde{x} \\ 0 & 1 \end{pmatrix} =$$

$$= \begin{pmatrix} x + \cdots & -x \ln \tilde{x} + \alpha x + \cdots \\ \beta x^3 + \cdots & 1/x + \cdots \end{pmatrix}. \tag{2.3.16}$$

Let us define (see [Bo2]) *the Fuchsian weight of a Fuchsian second order system (2.1.6) with singularities $a_1, \ldots, a_n$ as*

$$\gamma_{(B)} = \sum_{i=1}^{n} \left(\varphi_i^1 - \varphi_i^2 \right).$$

For our B (defined by (2.1.4)) $\gamma_{(B)} = 2$ (cf.(2.3.2) and (2.3.5)).

This notion is quite "constructive", as we know how to compute φ_i^j for a given Fuchsian system. The next definition will not be so constructive. For given "Hilbert data" ($\{a_i\}$ and a representation $\chi : \Delta \to GL(2, \mathbb{C})$, more precisely, a class $\{C^{-1}\chi C\}$ of mutually conjugate representations) consider $\min \gamma_{(D)}$ over all Fuchsian systems $\dot{z} = Dz$ having singularities a_i and monodromy χ (more precisely, $\{C^{-1}\chi C\}$). Let us denote this $\min \gamma_{(D)}$ by γ_χ and call it *the Fuchsian weight of representation* χ (without mentioning the a_i explicitly. However, since Δ is the group of deck transformations of the covering $\tilde{S} \to S$, where $S = \bar{\mathbb{C}} \setminus \{a_1, \ldots, a_n\}$, one can say that any Δ is always related to some a_i and so we "have them in mind" when we mention Δ, hence when we mention χ).

Lemma 2.3.1 *If a_i are the singularities of B in (2.1.4) and χ is the monodromy of (2.1.6) with this B, then $\gamma_\chi = 2$.*

So the problem

$$\left\{ \begin{array}{c} \text{"realize the above - mentioned } a_i \text{ and } \chi \text{ by a} \\ \text{Fuchsian system } \dot{y} = Dy \text{ satisfying the additional} \\ \text{requirement } \gamma_{(D)} < 2\text{"} \end{array} \right\} \qquad (2.3.17)$$

cannot be done. It is precisely the auxiliary problem mentioned in Section 2.1.

Assume that (2.3.17) can be done. For the corresponding system $\dot{y} = Dy$ we shall denote the numbers mentioned in (2.2.39) by $\mu_i^j(D)$, $\rho_i^j(D)$, $\varphi_i^j(D)$, $\beta_i^j(D)$. Four our χ the matrices $G_i = \chi(\sigma_i^{-1})$ are Jordan unipotent. Hence, $E_i = \frac{1}{2\pi i} \ln G_i$ are nilpotent of rank 1 and their eigenvalues $\mu_i^j(D) = 0$. Thus, $\rho_i^j(D) = 0$ and $\beta_i^j(D) = \varphi_i^j(D)$. Let $k := \sum_{i=1}^4 \varphi_i^1(D)$, $l := \sum_{i=1}^4 \varphi_i^2(D)$. Then $\gamma_{(D)} = k - l$. It is clear that $k, l \in \mathbb{Z}$ and $k \geq l$ (all $\varphi_i^1(D) \geq \varphi_i^2(D)$). But, according to Theorem 2.2.2,

$$k + l = \sum_{i,j} \varphi_i^j(D) = \sum_{i,j} \beta_i^j(D) = 0.$$

It follows that $k = -l$, $\gamma_{(D)} = 2k$, and if $0 \leq \gamma_{(D)} < 2$, then $k = 0$, $l = 0$, $\gamma_{(D)} = 0$. So all $\varphi_i^1(D) = \varphi_i^2(D)$.

We see that for a fixed a_i all solutions to $\dot{y} = Dy$ have the same growth at a_i (neglecting fractional powers and ln-s). But for different a_i the growth may be different. Now we shall modify our system so that the new system will have the same singularities and monodromy, but the valuation will become the same at all singular points (and of course, still for all of its solutions).

Let $m_i := \varphi_i^1(D) = \varphi_i^2(D)$. Introduce a new dependent variable $z = fy$, where $f = \prod_{i=1}^4 (x - a_i)^{-m_i}$. Evidently,

$$\dot{z} = f\dot{y} + \dot{f}y = \left(D + \dot{f}/f\right) fy = Fz, \qquad (2.3.18)$$

where $F = D + (\ln f)\dot{} I = D - \sum_{i=1}^4 \frac{m_i}{x-a_i} I$. If $D = \sum_{i=1}^4 \frac{1}{x-a_i} D_i$, then $F = \sum_{i=1}^4 \frac{1}{x-a_i} F_i$, where $F_i = D_i - m_i I$, so the system (2.3.18) is Fuchsian. It satisfies the condition analogous to (1.2.3): $\sum_{i=1}^4 F_i = 0$. Indeed, $\sum_{i=1}^4 D_i = 0$, since $\dot{y} = Dy$ has no singularity at ∞, and $\sum_{i=1}^4 m_i = k = 0$. Clearly $\dot{z} = Fz$ has the desired properties (for any of its solutions z we have $\varphi_i(z) = \varphi_i(y) - m_i = 0$). This means that for any matrix solution Z to $\dot{Z} = FZ$ one has

$$
\begin{aligned}
Z(\tilde{x}) &= W_i(x)(\tilde{x} - a_i)^{E_i}, \quad x \in \tilde{U}_i^*, \\
Z(\tilde{x}) &= W_i(x)(\tau^{-1}\tilde{x} - a_i)^{E_i}\chi(\tau^{-1}), \quad x \in \tau^{-1}\tilde{U}_i^*
\end{aligned}
\qquad (2.3.19)
$$

(cf. the discussion of (2.3.3), (2.3.4)). Of course, E_i and $\chi(\tau^{-1})$ depend somehow on Z. Let Y be a Levelt's fundamental system of solutions to (2.1.6) satisfying the additional requirement $V_1(0) = I$. Let χ be the corresponding monodromy representation (defined by this Y via (2.1.5)). Take a solution Z to $\dot{Z} = FZ$ having the same monodromy χ. In $p^{-1}(U_2^*)$, $p^{-1}(U_3^*)$, $p^{-1}(U_4^*)$ (2.3.4) holds, whereas in $p^{-1}(U_1^*)$

$$Y(\tilde{x}) = V_1(x)x^{\Phi}(\tau^{-1}\tilde{x})^{E_1}\chi(\tau^{-1}), \quad \text{if } x \in \tau U_i^*.$$

Consider YZ^{-1}. It has the following properties:

$$YZ^{-1} \text{ is holomorphic in } \tilde{S} \qquad (2.3.20)$$

$$YZ^{-1} \circ \sigma = (Y \circ \sigma)(Z \circ \sigma)^{-1} = Y\chi^{-1}(\sigma)\chi(\sigma)Z^{-1} = YZ^{-1}, \qquad (2.3.21)$$

$$YZ^{-1} = V_1(x)x^{\Phi}W_1(x)^{-1}, \text{ if } \tilde{x} \in p^{-1}(U_1^*) \qquad (2.3.22)$$

$$YZ^{-1} = V_1(x)W_1(x)^{-1}, \text{ if } \tilde{x} \in p^{-1}(U_i^*), \ i \geq 2. \qquad (2.3.23)$$

Note that in (2.3.22) the other factors cancel; (2.3.20) and (2.3.21) imply that YZ^{-1} is a holomorphic function on S. More precisely $(YZ^{-1})(\tilde{x})$ is the same for all $\tilde{x} \in p^{-1}(x)$, so $x \mapsto (YZ^{-1})(\tilde{x})$ is a well-defined holomorphic function on S. The property (2.3.23) together with $W_1(0) \neq 0$ (see Theorem 2.2.1) show that it is holomorphic in $U_2 \cup U_3 \cup U_4$. And in U_1 we have (cf. (2.3.15))

$$
YZ^{-1} + V_1(x)x^{\Phi}W_1^{-1}(x) =
$$
$$
= \begin{pmatrix} O(1) & O(x^2) \\ O(x) & O(1) \end{pmatrix} \begin{pmatrix} x & O \\ 0 & x^{-1} \end{pmatrix} \cdot \text{(a holomorphic matrix)} =
$$
$$
= \begin{pmatrix} O(x) & O(x) \\ * & * \end{pmatrix}.
$$

It follows that the first row of YZ^{-1} is holomorphic in U_1, so it is holomorphic in all of $\bar{\mathbb{C}}$, thus a constant. Being $O(x)$ in U_1, it must be identically 0. This contradicts the fact that YZ^{-1} is invertible away from the singularities.

2.4 The third order system

We are going to study the system (1.2.1) with A as in (2.1.2). It is equivalent to the system (2.1.6) with B as in (2.1.4) (the dependent variables in these systems now have numbers 2 and 3) together with the equation (2.1.5). Let us check that (1.2.1) has no singular point at ∞. This is already known for the quotient system (2.1.6), so we need only to rewrite (2.1.5) in terms of the independent variable $t = 1/x$. We get

$$
\frac{dy}{dt} = -\frac{1}{t^2}\left(a_{12}\left(\frac{1}{t}\right)y^2 + a_{13}\left(\frac{1}{t}\right)y^3\right).
$$

It is easily checked that $\frac{1}{t^2}a_{12}\left(\frac{1}{t}\right)$, $\frac{1}{t^2}a_{13}\left(\frac{1}{t}\right)$ have no pole at $t = 0$.

The singular points $a_2 = -1$, $a_3 = 1$, $a_4 = 1/2$ are Fuchsian. Near these points $A(x) = B_i(x)/(x - a_i)$ for some $B_i(x)$ holomorphic in U_i. So the corresponding $L_i(0)$ are similar to $B_i(0)$, which in this case are

$$
\frac{1}{6}\begin{pmatrix} 0 & 6 & 0 \\ 0 & -1 & 1 \\ 0 & -1 & 1 \end{pmatrix}, \quad \frac{1}{2}\begin{pmatrix} 0 & 0 & 2 \\ 0 & -1 & -1 \\ 0 & 1 & 1 \end{pmatrix}, \quad \frac{1}{3}\begin{pmatrix} 0 & -3 & -3 \\ 0 & -1 & 1 \\ 0 & -1 & 1 \end{pmatrix}.
$$

Under the transformation $y \mapsto B_i(0)y$ variables y^2, y^3 are transformed (independently of y^1) via (2.3.1). Hence under the transformation $y \mapsto B_i^2(0)y$ they become 0. Acting by $B_i(0)$ once more, we obtain $y^1 = 0$, as the new y^1 is some linear combination of the previous y^2, y^3. Thus all $B_i^3(0) = 0$, i.e., $B_i(0)$ is nilpotent. At the same time it is clear that the determinant of the second order block lying in the upper right corner of $B_i(0)$

$$
\begin{pmatrix}
\cdot & * & * \\
\cdot & * & * \\
\cdot & \cdot & \cdot
\end{pmatrix}
$$

is non-zero. So the $B_i(0)$ are of rank 2, and the $L_i(0)$ are also nilpotent matrices of rank 2. Each $L_i(0)$ consists of one diagonal block, consequently, $\Phi_i + E_i = L_i(0)$. This implies that $\Phi_i = 0$. Thus

$$\varphi_i^j = 0, \ E_i \text{ is nilpotent of rank 2,} \qquad (2.4.1)$$
$$\mu_i^j = 0, \ \rho_i^j = 0, \ \beta_i^j = 0, \ (i = 2, 3, 4; \ j = 1, 2, 3)$$

Now consider the more complicated singular point 0. If y^2 and y^3 have only polynomial growths as $x \to 0$, then so does the right hand side of (2.1.5), and y^1 (being the integral of the latter) also grows at most polynomially. This proves that 0 is a regular singularity of (1.2.1).

Now we claim that one can obtain a Levelt's basis for (1.2.1) at the point 0 in the following way. Firstly, take $y_1 = \begin{pmatrix} 1 \\ 0 \\ 0 \end{pmatrix}$. Secondly, let $\bar{y}_2 = \begin{pmatrix} y_2^2 \\ y_2^3 \end{pmatrix}$, $\bar{y}_3 = \begin{pmatrix} y_3^2 \\ y_3^3 \end{pmatrix}$ be a Levelt's basis for (2.1.6) (with B as in (2.1.4) and $V(0) = I$. We have increased the numbers of these vectors, by one, as with the numbers of coordinates). Starting from these y_i^j, we complete the construction of the full solutions y_2, y_3 to (1.2.1) just using (2.1.5):

$$
\begin{aligned}
y_2^1 &= \int (a_{12}y_2^2 + a_{13}y_2^3)dx, \\
y_3^1 &= \int (a_{12}y_3^2 + a_{13}y_3^3)dx
\end{aligned}
$$

(regardless of the constants of integration). From (2.1.3) and (2.3.16) we have

$$a_{12}y_2^2 + a_{13}y_2^3 = \frac{1}{x^2}(x + \cdots) + (\text{hol.})(x + \cdots) + (\text{hol.})(\beta x^3 + \cdots) =$$

$$= \frac{1}{x} + \text{hol.}$$

where "hol." means holomorphic. (Since $\bar{y}_2$ does not "branch" near 0, i.e., $\bar{y}_2 \circ \sigma_1 = \bar{y}_2$, the omitted terms here are not only of higher order, but they are also series in integral powers of x). Hence

$$y_2^1 = \ln \tilde{x} + f, \quad \text{where } f \text{ is holomorphic in } U_1 \text{ and } \varphi(f) > 0. \qquad (2.4.2)$$

So

$$\lambda y_1 + \mu y_2 = \begin{pmatrix} \lambda + \mu \ln \tilde{x} + \text{hol.} \\ \text{hol.} \\ \text{hol.} \end{pmatrix}$$

Clearly for all $(\lambda, \mu) \neq (0, 0)$

$$\varphi(\lambda y_1^1 + \mu y_2^1) = \varphi(\lambda + \mu \ln \tilde{x} + f) = 0$$

(the constant and logarithm terms can never cancel). For the other coordinates $\varphi \geq 0$ (in fact, > 0). Thus we have found a two-dimensional subspace $X_1^1 = \mathbb{C}y_1 \oplus \mathbb{C}y_2$ in X_1 such that $\varphi(X_1^1 \setminus 0) = 0$.

With regard to y_3, it turns out that $\varphi(y_3) = -1$. Indeed, $a_{13}(0) = -\frac{1}{2}$,

$$a_{12}y_3^2 + a_{13}y_3^3 =$$

$$= \left(\frac{1}{x^2} + \text{hol.} \right)(-x \ln \tilde{x} + \alpha x + \cdots) + \left(-\frac{1}{2} + \cdots \right)\left(\frac{1}{x} + \cdots \right) =$$

$$= -\frac{\ln \tilde{x}}{x} + \left(\alpha - \frac{1}{2} \right)\frac{1}{x} + \cdots,$$

$$y_3^1 = -\frac{1}{2}\ln^2 \tilde{x} + \left(\alpha - \frac{1}{2} \right)\ln \tilde{x} + \cdots.$$

Clearly $\varphi(y_3^1) = 0$, $\varphi(y_3^2) = 0$, $\varphi(y_3^3) = -1$, and the minimum of these is -1.

This completes the determination of the filtration (2.2.9) and the numbers φ_1^i: as $X_1^1 \oplus \mathbb{C}y_3 = X$ the filtration (2.2.9) in our case is

$$0 = X^0 \subset X_1^1 \subset X_1^2 = X_1,$$

and $\varphi_1^1 = \varphi_1^2 = 0$, $\varphi_1^3 = -1$.

Now let us inquire about E_1. Clearly $\sigma_1^* y_1 = y_1 \circ \sigma_1 = y_1$, and (2.4.2) and $\bar{y}^2 \circ \sigma_1 = \bar{y}^2$ imply that $\sigma_1^* y_2 = y_2 \circ \sigma_1 = y_2 + 2\pi i y_1$. More generally, consider $y_2 \circ \sigma$ for any $\sigma \in \Delta$. Clearly $\ln \circ \sigma = \ln + \chi_1(\sigma^{-1})$ where χ_1 is some representation $\Delta \to 2\pi i \mathbb{Z}$ (the latter group being additive). It follows from (2.4.2) that $y_2 \circ \sigma = \chi(\sigma^{-1}) y_1 + y_2$. Of course $y_1 \circ \sigma = y_1$. Now, writing Y in a somewhat condensed form

$$Y = \begin{pmatrix} 1 & y_2^1 & y_3^1 \\ 0 & \bar{y}_2 & \bar{y}_3 \end{pmatrix}$$

(the entries of the second row are column vectors with 2 entries), we have

$$Y\chi(\sigma^{-1}) = Y \circ \sigma = \begin{pmatrix} 1 & y_2^1 \circ \sigma & y_3^1 \circ \sigma \\ 0 & \bar{y}_2 \circ \sigma & \bar{y}_3 \circ \sigma \end{pmatrix} =$$

$$= \begin{pmatrix} 1 & y_2^1 & y_3^1 \\ 0 & \bar{y}_2 & \bar{y}_3 \end{pmatrix} \begin{pmatrix} 1 & \chi_1(\sigma^{-1}) & a \\ 0 & 1 & b \\ 0 & 0 & c \end{pmatrix}$$

for some a, b, c. This implies

$$(\bar{y}_2, \bar{y}_3) \circ \sigma = (\bar{y}_2, \bar{y}_3) \begin{pmatrix} 1 & b \\ 0 & c \end{pmatrix}.$$

But we know that

$$(\bar{y}_2, \bar{y}_3) \circ \sigma = (\bar{y}_2, \bar{y}_3) \bar{\chi}(\sigma^{-1}),$$

where $\bar{\chi}$ is a monodromy representation $\Delta \to GL(2, \mathbb{C})$ which corresponds to the fundamental system of solutions $(\bar{y}_2, \bar{y}_3)$ to (2.1.6). Hence replacing σ^{-1} by σ we get

$$\chi(\sigma) = \begin{pmatrix} 1 & \chi_1(\sigma) & * \\ 0 & & \bar{\chi}(\sigma) \end{pmatrix} \qquad (2.4.3)$$

In the particular case $\sigma = \sigma_1$ we see that $Y \circ \sigma_1 = Y G_1$, where

$$G_1 = \begin{pmatrix} 1 & 2\pi i & * \\ 0 & 1 & -2\pi i \\ 0 & 0 & 1 \end{pmatrix}.$$

$(\chi_1 (\sigma_1^{-1}) = 2\pi i$ follows directly from the definition of χ_1, and

$$\bar{\chi}(\sigma_1) = e^{2\pi i E_1} = I + 2\pi i \bar{E}_1 = \begin{pmatrix} 1 & -2\pi i \\ 0 & 0 \end{pmatrix},$$

where $\bar{E}_1$ is the matrix which was denoted by E_1 in Section 2.3). It follows, firstly, that $E_1 = \frac{1}{2\pi i} \ln G_1$ is a nilpotent matrix. This implies that $E_1^3 = 0$ and E_1^2 is of the form $\begin{pmatrix} 0 & 0 & * \\ 0 & 0 & 0 \\ 0 & 0 & 0 \end{pmatrix}$, so

$$e^{2\pi i E_1} = I + 2\pi i E_1 + \begin{pmatrix} 0 & 0 & * \\ 0 & 0 & 0 \\ 0 & 0 & 0 \end{pmatrix}.$$

Now we see, secondly, that

$$E_1 = \begin{pmatrix} 0 & 1 & * \\ 0 & 0 & -1 \\ 0 & 0 & 0 \end{pmatrix},$$

so E_1 has rank 2.

Collecting together the properties of our system at 0:

$$\varphi_1^1 = \varphi_1^2 = 0, \ \varphi_1^3 = -1, \ E_1 = \text{ is nilpotent of rank 2,} \qquad (2.4.4)$$
$$\mu_1^j = 0, \ \rho_1^j = 0, \ \beta_1^j = \varphi_1^j \ (j = 1, 2, 3).$$

It is important that linear transformations σ_i^* $(i = 1, 2, 3, 4)$ in X have the same eigenspace. It is the 1-dimensional vector space ("a line") $\mathbb{C}y_1 =$

$= \mathbb{C} \begin{pmatrix} 1 \\ 0 \\ 0 \end{pmatrix}$. Indeed, for each σ_i^* the eigenspace is 1-dimensional, as σ_i^* has a matrix

representation $e^{2\pi i E_i}$ in some basis (which may depend on i) and E_i is nilpotent of rank 2. But $\sigma_i^* y_1 = y_1 \circ \sigma_i = y_1$, so y_1 belongs to all of these 1-dimensional subspaces; hence they all coincide with $\mathbb{C}y_1$.

At last we are in a position to prove the Assertion made in Section 2.1. Assume that $\dot{z} = Cz$ is a Fuchsian system having the same singularities and monodromy as (1.2.1). Let $\tilde{X}$ be the space of its solutions. Let $Y = (y_1, y_2, y_3)$ be an invertible matrix solution to (1.2.4) and let χ be the corresponding monodromy representation. The matrix system $\dot{Z} = CZ$ has an invertible solution with the same monodromy representation; let $Z = (z_1, z_2, z_3)$ be this solution. Define the linear map $\theta : X \to \tilde{X}$ in such a way that $\theta\,_i = z_i$, $i = 1, 2, 3$. It is an isomorphism commuting with the right action of Δ on these spaces (the action is $(y, \sigma) \mapsto y \circ \sigma$, $(z, \sigma) \mapsto z \circ \sigma$). Indeed, any vector $y \in X$ can be written as $Y\eta$, where η is some column (η^1, η^2, η^3). Then θ maps y to $z = Z\eta$, and $y \circ \sigma = Y\eta \circ \sigma = (Y \circ \sigma)\eta = Y\chi^{-1}(\sigma)\eta$ into $Z\chi^{-1}(\sigma)\eta = (Z \circ \sigma)\eta = Z\eta \circ \sigma = z \circ \sigma$.

It follows that $z_1 = \theta \begin{pmatrix} 1 \\ 0 \\ 0 \end{pmatrix}$ is a common eigenvector of $\sigma_1^*, \sigma_2^*, \sigma_3^*, \sigma_4^*$, i.e.,

$z_1 \circ \sigma_i = z_1$. This means that the vector z_1 is a single-valued holomorphic vector function on S. Having only polynomial growth near singular points a_i, it is a meromorphic vector function on $\bar{\mathbb{C}}$. If it were a scalar meromorphic function, then we would have $\sum_{i=1}^{4} \varphi_i(y_1) = 0$ ("the sum of the orders of the zeros equals the sum of the orders of the poles"). But for a meromorphic vector function one can only state that $\sum_{i=1}^{4} \varphi_i(y) \leq 0$. Indeed,

$$\sum_{i=1}^{4} \varphi_i(y) = \sum_{i=1}^{4} \min\{\varphi_i(y^j), j = 1, 2, 3\}.$$

But for any k, $\min\{\varphi_i(y^j), j = 1, 2, 3\} \leq \varphi_i(y^k)$, so

$$\sum_{i=1}^{4} \varphi_i(y) \leq \min\left\{\sum_{i=1}^{4} \min \varphi_i(y^k), k = 1, 2, 3\right\} = 0.$$

At the same time, $\sigma_i^* z_1 = z_1$ implies that z_1 is the first vector of some Levelt's basis at a_i. So $\varphi_i(z_1)$ is the maximum value taken by φ_i, i.e., $\varphi_i(z_1) = \varphi_i^1(C)$. Let $k := \sum_{i=1}^{4} \varphi_i^1(C)$, $l := \sum_{i=1}^{4} \varphi_i^2(C)$, $m := \sum_{i=1}^{4} \varphi_i^3(C)$. We have just seen that $k \leq 0$. But $k \geq l$ (since all $\varphi_i^1(C) \geq \varphi_i^2(C)$), $l \geq m$ and $k + l + m = 0$. It follows that $k = 0$, $l = 0$, $m = 0$. Furthermore, $\varphi_i^1(C) = \varphi_i^2(C) = \varphi_i^3(C)$ for all i.

The next step is analogous to the step made in Section 2.3. Let $n_i = \varphi_i^1(C)$. Introduce a new dependent variable $w = fz$, $f = \prod_{i=1}^{4}(x - a_i)^{-n_i}$. We arrive at the Fuchsian system $\dot{w} = Dw$ having the same singularities and monodromy as (1.2.1) and possessing the property that all $\varphi_i^j = 0$. Now $w_1 = fz_1$ is a single-valued function holomorphic on $\bar{\mathbb{C}}$. Hence $w_1 =$const. There exists an invertible matrix M

such that $w_1 = M \begin{pmatrix} 1 \\ 0 \\ 0 \end{pmatrix}$. Introduce a new dependent variable $u = M^{-1}w$. If w satisfies $\dot{w} = Dw$, u satisfies the system

$$\dot{u} = M^{-1}DMu \qquad (2.4.5)$$

which is still Fuchsian with singularities a_i and monodromy χ. (The monodromy describes how $w_j \circ \sigma$ can be represented as a linear combination of w_1, w_2, w_3; but then the same is valid for $u_j = M^{-1}w_j$). Of course $\varphi_i^j(M^{-1}DM) = 0$, i.e., $\varphi_i(u) = 0$ for any solution $u \neq 0$ to (2.4.5). The system (2.4.5) has a solution $u_1 = M^{-1}w_1 = \begin{pmatrix} 1 \\ 0 \\ 0 \end{pmatrix}$, hence the first column of $M^{-1}DM$ is 0 and

$$M^{-1}DM = \begin{pmatrix} 0 & * \\ 0 & F \\ 0 & \end{pmatrix}$$

for some matrix $F(x)$. Let us consider the system

$$\dot{v} = Fv, \quad v \in \mathbb{C}^2 \qquad (2.4.6)$$

which, of course is Fuchsian with singularities $a_1, \dots, a_4$. We know that

$$(u_1, u_2, u_3) \circ \sigma = (u_1, u_2, u_3)\chi\left(\sigma^{-1}\right)$$

where $\chi\left(\sigma^{-1}\right)$ has the form (2.4.3). Let

$$u_2 = \begin{pmatrix} u_2^1 \\ \bar{u}_2 \end{pmatrix}, \quad u_3 = \begin{pmatrix} u_3^1 \\ \bar{u}_3 \end{pmatrix},$$

where the $\bar{u}_i$ themselves are 2-columns. They are solutions to (2.4.6) and are clearly independent, since the matrix

$$(u_1, u_2, u_3) = \begin{pmatrix} 1 & u_2^1 & u_3^1 \\ 0 & \bar{u}_2 & \bar{u}_3 \end{pmatrix}$$

would otherwise be degenerate.

We conclude from (2.4.3) that

$$(\bar{u}_2, \bar{u}_3) \circ \sigma = (\bar{u}_2, \bar{u}_3)\bar{\chi}\left(\sigma^{-1}\right).$$

In other words, $(\bar{u}_2, \bar{u}_3)$ is a fundamental system of solutions to (2.4.6) and the corresponding monodromy is the same as for the system (2.1.6).

Our final step is to prove that the Fuchsian weight

$$\gamma_{(F)} = 0 < \gamma_{(B)} \tag{2.4.7}$$

which contradicts to the Lemma from Section 2.3. Any solution v to (2.4.6) can be considered as a "subcolumn" of some vector function u which is a solution to (2.4.5). So $v = \begin{pmatrix} u^2 \\ u^3 \end{pmatrix}$,

$$\varphi_i(v) = \min(\varphi_i(u^2), \varphi_i(u^3)) \geq$$
$$\geq \min(\varphi_i(u^1), \varphi_i(u^2), \varphi_i(u^3)) = \varphi_i(u) = 0.$$

All the $\varphi_i^j(F)$ are $\varphi_i(v)$ for some v, hence

$$\text{all } \varphi_i^j(F) \geq 0. \tag{2.4.8}$$

The system (2.4.6) has the same monodromy as (2.1.6), so all the $\mu_i^j(F) = 0$, $\rho_i^j(F) = 0$, $\beta_i^j(F) = \varphi_i^j(F)$. Theorem 2.2.2 implies that $\sum_{i,j} \varphi_i^j(F) = 0$. According to (2.4.8), this may happen only if all $\varphi_i^j(F) = 0$. Then we get (2.4.7):

$$\gamma_{(F)} = \sum_{i=1}^{4} (\varphi_i^1(F) - \varphi_i^2(F)) = 0.$$

3 The Plemelj theorem

3.1 A weak version of Plemelj's theorem

Some ideas which are essential for the proof of Plemelj's theorem manifest themselves clearly during the proof of the following weaker theorem: for any $a_1, \ldots, a_n$ and χ there exists a system (1.2.1) which is holomorphic on S and has monodromy representation χ. (It is not claimed here that the system is regular).

Assume we have constructed a holomorphic matrix-valued function $Y : \hat{S} \to GL(p, \mathbb{C})$ satisfying (1.2.5). Then we are through. Indeed, take

$$A := (dY(\tilde{x})/dx)Y^{-1}(x). \tag{3.1.1}$$

Clearly (1.2.5) implies

$$dY(\sigma\tilde{x})/dx = (dY(\tilde{x})/dx)\chi\left(\sigma^{-1}\right),$$

and $A(\sigma\tilde{x}) = A(\tilde{x})$. Thus A can be considered as a (single-valued) holomorphic function on S and Y is a nondegenerate matrix solution to (1.2.4) with this A. So this system really has a solution (namely, Y) with the demanded "branching" property (1.2.5).

In order to get Y we shall have to consider a somewhat different object – a "branching cross-section" Z of some principal bundle $P \to S$ with the standard fibre $GL(p, \mathbb{C})$; this Z will have the same branching property as Y. First we shall describe this principal bundle P as well as the corresponding vector bundle E with the standard fibre $\mathbb{C}^p$.

The universal covering $p : \tilde{S} \to S$ can be considered as a principal bundle over S with the standard fibre Δ (the group of deck transformations). This has been already explained in detail in Steenrod's classical book ([St]) and is well-known.

The only point here which needs some explanation is the following. We have a left action of Δ on $\tilde{S}$:

$$\Delta \times \tilde{S} \to \tilde{S} \quad (\sigma, \tilde{x}) \mapsto \sigma\tilde{x},$$

whereas for the principal bundles it is standard to have a right action of the structural group (= standard fibre) on the total space. Of course we can simply denote $\sigma\tilde{x}$ by $\tilde{x}\sigma$ which essentially means that the former product $\sigma\tau$ in Δ will now be denoted as $\tau\sigma$. But this would lead to another inconsistency – inconsistency with the standard construction of $\tilde{S}$ using classes of paths in S which also leads quite naturally to the identification of Δ with the fundamental group of S. (Multiplication in the latter group is defined according to the generally accepted agreement for the multiplication of paths – the product $\alpha\beta$ of the paths $\alpha, \beta : [0,1] \rightarrow S$ is defined if and only if $\alpha(1) = \beta(0)$; in this case, when t runs through $[0,1]$, $(\alpha\beta)(t)$ first runs along α and then along β). Thus there are sufficiently good reasons to regard the action of Δ on $\tilde{S}$ as a left action. However, with any left action

$$\Delta \times \tilde{S} \rightarrow \tilde{S} \quad (\sigma, \tilde{x}) \mapsto \sigma\tilde{x},$$

one can associate a right action

$$\tilde{S} \times \Delta \rightarrow \tilde{S} \quad (\tilde{x}, \sigma) \mapsto \tilde{x} \cdot \sigma$$

just defining it by a formula $\tilde{x} \cdot \sigma = \sigma^{-1}\tilde{x}$. It is really a right action:

$$\tilde{x} \cdot (\sigma\tau) = (\sigma\tau)^{-1}\tilde{x} = \tau^{-1}\sigma^{-1}\tilde{x} = (\sigma^{-1}\tilde{x}) \cdot \tau = (\tilde{x} \cdot \sigma) \cdot \tau.$$

Having this right action in mind, we can consider $p : \tilde{S} \rightarrow S$ as a principal bundle with the structural group Δ.

Linear representation $\chi : \Delta \rightarrow GL(p, \mathbb{C})$ defines a left action of Δ in $\mathbb{C}^p$ and $GL(p, \mathbb{C})$. (On $GL(p, \mathbb{C})$ we set $(\sigma, Y) \mapsto \chi(\sigma)Y$ using the multiplication of matrices). Now we can construct the bundles E, P which are the bundles associated to $\tilde{S}$ with the standard fibres $\mathbb{C}^p$, $GL(p, \mathbb{C})$ and with the above mentioned action of Δ in these standard fibres. We recall that they can be obtained as follows:

$$E = \tilde{S} \times \mathbb{C}^p / \sim, \quad P = \tilde{S} \times GL(p, \mathbb{C}) / \sim, \tag{3.1.2}$$

where for E

$$(\tilde{x}, y) \sim (\tilde{x} \cdot \sigma, \chi(\sigma^{-1})\, y) = (\sigma^{-1}\tilde{x}, \chi(\sigma^{-1})\, y),$$

in other words,

$$(\tilde{x}, y) \sim (\sigma\tilde{x}, \chi(\sigma)y). \tag{3.1.3}$$

Analogously for P

$$(\tilde{x}, Y) \sim (\sigma\tilde{x}, \chi(\sigma)Y). \tag{3.1.4}$$

Denote projections $E \to S, P \to S$ by p_E, p_P.

Although this is not absolutely necessary, we recall that there exists also a "coordinate-wise" description of E, P which sometimes seems to be more "pictorial" (however, the "real work" is done in terms of (3.1.2), (3.1.3), (3.1.4)). S is covered by small neighborhoods V_α. Over V_α in $\tilde{S}$ there exists a local cross-section $\tilde{x}_\alpha : V_\alpha \to p^{-1}(V_\alpha)$ which allows to define "local coordinates" over V_α: if $\tilde{x} \in p^{-1}(V_\alpha)$, $p\tilde{x} = x$, then $\tilde{x} = \xi\tilde{x}_\alpha(x)$ with some (uniquely determined) $\xi \in \Delta$, and we could set $\Phi_\alpha(\tilde{x}) = (x, \xi)$. However, in order to be consistent in treating $\tilde{S}$ as a space with the right Δ-action, we shall set $\Phi_\alpha(\tilde{x}) = (x, \xi^{-1})$. These coordinates are equivariant with respect to the right action of $\Delta : \Phi_\alpha(\tilde{x}\sigma) = (x, \xi\sigma)$.

If for some bundle we have (using the same notation)

$$\tilde{x} \in V_\alpha \cap V_\beta, \quad \Phi_\alpha(\tilde{x}) = (x, \xi), \quad \Phi_\beta(\tilde{x}) = (x, \eta),$$

then there arises a coordinate transformation $g_{\alpha\beta}(x)$:

$$\xi = g_{\alpha\beta}(x)\eta.$$

(This means that $\tilde{x}_\alpha(x) = g_{\alpha\beta}\tilde{x}_\beta(x)$.) For $p : \tilde{S} \to S$ these $g_{\alpha\beta}(x)$ are locally constant (in x), and we may even assume that they are constant (e.g., this will happen if all $V_\alpha \cap V_\beta$ are connected). For the associated bundles E, P we have local coordinates

$$\Phi_\alpha^E : p_E^{-1}(V_\alpha) \to V_\alpha \times \mathbb{C}^p, \quad \Phi_\alpha^P : p_P^{-1}(V_\alpha) \to V_\alpha \times GL(p, \mathbb{C})$$

which are defined as follows in terms of (3.1.2)–(3.1.4) and $\tilde{x}_\alpha$. Any point of $p_E^{-1}(V_\alpha)$ or $p_P^{-1}(V_\alpha)$ is a class of equivalence containing just one element of the form

$$(\tilde{x}_\alpha(x), y) \text{ or } (\tilde{x}_\alpha(x), Y),$$

where $y \in \mathbb{C}^p, Y \in GL(p, \mathbb{C})$. Then

$$\Phi_\alpha^E \langle (\tilde{x}_\alpha(x), y) \rangle = (x, y), \quad \Phi_\alpha^P \langle (\tilde{x}_\alpha(x), Y) \rangle = Y$$

($\langle \rangle$ means the corresponding class of equivalence). For E the coordinate transformations are linear transformations $\chi(g_{\alpha\beta})$, and for P they are left shifts of $GL(p, \mathbb{C})$ on $\chi(g_{\alpha\beta})$. These transformations and shifts are constant (in $V_\alpha \cap V_\beta$).

This special structure in our bundles (they are bundles with the discrete structural group) allows one to speak "locally" (say, over V_α) about the "horizontal" cross-sections $V_\alpha \rightarrow P$ (and $V_\alpha \rightarrow E$ as well, but we need only the former). Namely, such are the local sections $Z : V_\alpha \rightarrow P$ for which $p_2 \Phi_\alpha^P(Z(x)) =$const, where p_2 is the standard projection

$$p_2 : V_\alpha \times GL(p, \mathbb{C}) \rightarrow GL(p, \mathbb{C}), \quad (x, Y) \mapsto Y.$$

If we have such a Z over V_α and $V_\alpha \cap V_\beta \neq \emptyset$, we can define a "horizontal continuation" of this section over V_β. Let $p_2 \Phi_\alpha^P(Z) = Y$; we define for all $x \in V_\beta$

$$Z(x) := \left(\Phi_\beta^P\right)^{-1} \left(x, \chi\left(g_{\alpha\beta}^{-1}\right) Y\right).$$

It is clear that this process can be continued, like the process of analytic continuation of the holomorphic function. However, generally it also gives us a "multivalued section" Z which can be considered as a single-valued map $Z : \tilde{S} \rightarrow P$ (having the property that $p_P Z = p$).

Instead of a more careful elaborating this pictorial idea, we shall define a required map $Z : \tilde{S} \rightarrow P$ using (3.1.2), (3.1.4). We simply set $Z(\tilde{x}) = \langle(\tilde{x}, I)\rangle$, where I is the identity matrix of p-th order. Then

$$Z(\sigma\tilde{x}) = \langle(\sigma\tilde{x}, I)\rangle = \left\langle\left(\tilde{x}, \chi\left(\sigma\right)^{-1}\right)\right\rangle = \langle(\tilde{x}, I), \chi\left(\sigma^{-1}\right)\rangle =$$
$$= \langle(\tilde{x}, I)\rangle\chi\left(\sigma^{-1}\right) = Z(\tilde{x})\chi\left(\sigma^{-1}\right), \qquad (3.1.5)$$
$$Z(\tilde{x}) = Z(\sigma\tilde{x})\chi(\sigma).$$

Thus our Z really has the same branching properties which are required from Y.

Now let us enlarge the structural group of E, P from $\chi(\Delta)$ to $GL(p, \mathbb{C})$. This means that we consider P as a principal $GL(p, \mathbb{C})$-bundle and E as a vector bundle associated to it. In terms of (3.1.2)–(3.1.4) the right action of $GL(p, \mathbb{C})$ on P arises from the action of this group on the corresponding direct product, so that the element $Z \in GL(p, \mathbb{C})$ acts as follows:

$$\langle(x, Y)\rangle Z = \langle(x, YZ)\rangle.$$

This definition is correct, since in $\tilde{S} \times GL(p, \mathbb{C})$

$$(\sigma\tilde{x}, \chi(\sigma)Y) \mapsto (\sigma\tilde{x}, \chi(\sigma)YZ) \sim (\tilde{x}, YZ).$$

An element of $p_P^{-1}(x)$, $x \in S$, can be interpreted as a basis $Y = (y_1, \ldots, y_p)$ in the vector space $E(x) := p_E^{-1}(x)$. By the way, the parametrization of $E(x)$ provided by such a basis can be expressed in the following way: Y can be interpreted as a map

$$Y : \mathbb{C}^p \to E(x), \quad Yz = \sum_{i=1}^{p} y_i z_i$$

(the latter is consistent with the record of Y as a "row" $(y_1, \ldots, y_p)$ and Z as a column with entries z_i). Note that the charts $(p_E^{-1}(V_\alpha), \Phi_\alpha^E)$ in E, $(p_P^{-1}(V_\alpha), \Phi_\alpha^P)$ in P introduce the complex-analytic structure in the total spaces of these bundles and that the bundles E, P are holomorphic with respect to this structure. Of course $\tilde{S}$ also has the standard complex-analytic structure defined by the charts $(p^{-1}(V_\alpha), p|p^{-1}(V_\alpha))$. In terms of these charts in S and the above mentioned charts in P the map $Z : \tilde{S} \to P$ has the local representation $x \mapsto (x, \text{const})$. Hence it is holomorphic.

We shall see in a moment that our bundles E, P (endowed with the structure of complex analytic vector, resp. principal bundle) are holomorphically equivalent to the direct products $S \times \mathbb{C}^p$, $S \times GL(p, \mathbb{C})$. Then there exists a ("true") holomorphic cross-section $W : S \to P$. Any element of $p_P^{-1}(x)$ can be obtained from $W(x)$ by multiplying it to the right by some matrix; in particular, $Z(\tilde{x}) = W(x)Y(\tilde{x})$, where $Y : \tilde{S} \to GL(p, \mathbb{C})$ is holomorphic. Clearly (3.1.5) implies that $Y(\tilde{x}) = Y(\sigma\tilde{x})\chi(\sigma)$, and we are through. S is homotopically equivalent (via an evident contraction) to a finite one-dimensional complex. The only topological obstruction for a real vector bundle over such a complex (and hence over S) to be nontrivial (nonequivalent to the direct product) can be its nonorientability. But any complex linear automorphism $\mathbb{C}^p \to \mathbb{C}^p$ considered as a map $\mathbb{R}^{2p} \to \mathbb{R}^{2p}$ has a positive determinant. So any complex vector bundle over S is trivial as a real vector bundle. Finally, S is a Stein manifold, and thus the topological triviality of E implies its holomorphic triviality, which means also the holomorphic triviality of P (see for details [Fö]).

The idea of this section is essentially the following. Let $F = S \times \mathbb{C}^p$, $Q = S \times GL(p, \mathbb{C})$. They are bundles over S with projections p_F, p_Q which are the standard projections on the first factor. Of course these bundles are trivial, but the systems (1.2.1), (1.2.4) allow one to define in F, Q the nontrivial structure of bundles with the discrete transformation group Δ (or $\chi(\Delta)$) and the same projections. (In the book [St] there is a section devoted to bundles with such a structure). We shall note three characteristic features of this structure (in particular, any of them defines this structure).

1) Considered with such a structure, these bundles must be associated with the principal Δ-bundle $\tilde{S} \to S$; and indeed, the corresponding identifications

$$F = \tilde{S} \times \mathbb{C}^p / \sim, \quad Q = \tilde{S} \times GL(p, \mathbb{C}) / \sim,$$

are given by the maps

$$(\tilde{x}, y) \mapsto Y(\tilde{x})y, \quad (\tilde{x}, Y) \mapsto Y(x)Y,$$

where $Y(\tilde{x})$ is a fixed nondegenerate solution of (1.2.4) satisfying (1.2.5).

2) One can also describe this structure in terms of suitable local coordinates in F, Q. Cover S by neighborhoods V_α such that (1.2.1), (1.2.4) have single-valued solutions there. Let $Y_\alpha(x)$ be any of the single-valued branches of Y in V_α. Local coordinates Φ_α^F, Φ_α^Q over V_α are obtained by inverting the parametrizations

$$V_\alpha \times \mathbb{C}^p \to p_F^{-1}(V_\alpha), \quad V_\alpha \times GL(p, \mathbb{C}) \to p_Q^{-1}(V_\alpha)$$
$$(x, y) \mapsto Y_\alpha(x)y, \quad (x, Y) \mapsto Y_\alpha(x)Y.$$

3) Finally, one can also define a continuation of solutions of (1.2.1), (1.2.4) along any path $\gamma : [0, 1] \to S$; this continuation is uniquely defined by its initial value. In terms of Φ_α^F, Φ_α^Q, it is a "lift" $u : [0, 1] \to F, U : [0, 1] \to Q, (p_F u = p_Q U = \gamma)$ which starts at a given point over $\gamma(0)$ and is "locally horizontal" in a sense that $p_2 \Phi_\alpha^F(u(t))$ =const, $p_2 \Phi_\alpha^Q(U(t))$ =const, while t runs over an interval J such that $\gamma(J) \subset V_\alpha$. Having all this in mind, we try to "reconstruct" F, Q without knowing $Y(\tilde{x})$. We first construct some "abstract" bundles E, P having the structure of the same type and then identify them with $S \times \mathbb{C}^p$, $S \times GL(p, \mathbb{C})$ using the general arguments from the previous paragraph. (When we speak about the existence of the cross-section $W : S \to P$, it is just another way to speak about such an identification). Thus our "branching cross-section" $Z(\tilde{x})$ becomes the required $Y(\tilde{x})$.

3.2   Proof of Plemelj's theorem

The defect of the simple arguments of the previous section is that they cannot guarantee that Y has (at most) polynomial growth near a and so (as we have already warned) they cannot guarantee that the corresponding system is regular. This is because the general reference to the theory of Stein manifolds is the last paragraph of the proof (the paragraph preceding "The idea of this section. . . ") provides no information on W except its existence. In this section we shall use a much more special theory which will enable us to control the behavior of some W near a_i (at the present moment we do not even know what this means). In regards to Z, it is a rather concrete object and so it is easy to control its growth near a_i (if we are able at all to prescribe an exact meaning to these words). Then it will follow that the growth of Y can be only polynomial, so that the system (1.2.1) with A from (3.1.1) is regular.

Perhaps the simplest situation which allows us to speak about the "growth of W near a_i" is when E, P are parts of bigger bundles G, R over the whole $\bar{\mathbb{C}}$ with projections p_G, p_R. Then a_i is contained in some small coordinate neighborhood U_i (to make things more definite let U_i be open disks centered at a_i and let the sizes of these U_i be the same and such that they do not intersect each other. We shall not consider the case $a_i = \infty$, because our goal is the Plemelj's theorem and it is sufficient to prove it assuming that all $a_i \neq \infty$). Over U_i we have a local coordinates

$$\Phi_i^G : p_G^{-1}(U_i) \to U_i \times \mathbb{C}^p, \ \Phi_i^R : p_R^{-1}(U_i) \to U_i \times GL(p, \mathbb{C}).$$

$W(x)$ is defined for all $x \in U_i^* := U_i \setminus \{a_i\}$ and we can speak about the growth of $p_2 \Phi_i^R(W(x))$ when $x \to a_i$ (here p_2 is again the standard projection on the second factor). We shall come into this situation by glueing E, P with $U_i \times \mathbb{C}^p$, $U_i \times GL(p, \mathbb{C})$ in an appropriate "fibrewise" way. This glueing will be defined by some maps

$$\omega_i^E : p_E^{-1}(U_i^*) \to U_i^* \times \mathbb{C}^p, \ \omega_i^P : p_P^{-1}(U_i^*) \to U_i^* \times GL(p, \mathbb{C}), \qquad (3.2.1)$$

which will preserve both the projection on U_i^* and the structure (of the vector space or of the right $GL(p, \mathbb{C})$-space) in the fibres.

Maps (3.2.1) will be defined in several steps.

Let again $G_i := \chi(\sigma_i^{-1})$, $E_i := \frac{1}{2\pi i} \ln G_i$ (with eigenvalues satisfying (2.2.2)). Define

$$f_i : \tilde{U}_i^* \times \Delta \to GL(p, \mathbb{C}) \quad f_i(\tilde{x}, \sigma) := (\tilde{x} - a_i)^{E_i} \chi(\sigma^{-1}).$$

Then

$$f_i \left(\sigma_i^k \tilde{x}, \sigma \sigma_i^{-k} \right) = f_i(\tilde{x}, \sigma). \qquad (3.2.2)$$

Really, it follows from (2.2.46) that

$$E_i \ln \left(\sigma_i^k \tilde{x} - a_i \right) = E_i \ln(\tilde{x} - a_i) + 2\pi i k E_i,$$
$$\left(\sigma_i^k \tilde{x} - a_i \right)^{E_i} = (\tilde{x} - a_i)^{E_i} \exp 2\pi i k E_i = (\tilde{x} - a_i)^{E_i} G_i^k,$$
$$\left(\sigma_i^k \tilde{x} - a_i \right)^{E_i} \chi \left(\left(\sigma \sigma_i^{-k} \right)^{-1} \right) = (\tilde{x} - a_i)^{E_i} G_i^k \chi \left(\sigma_i^k \right) \chi \left(\sigma^{-1} \right) = (\tilde{x} - a_i)^{E_i} \chi \left(\sigma^{-1} \right).$$

(3.2.2) allows one to define

$$g_i : p^{-1}(U_i^*) \to GL(p, \mathbb{C})$$

as follows: if $\tilde{x} \in p^{-1}(U_i^*)$, then $\tilde{x} = \sigma \tilde{x}_0$, $\tilde{x}_0 \in \tilde{U}_i^*$; take $g_i(\tilde{x}) := f_i(\tilde{x}_0, \sigma)$. We must check that this definition is correct, i.e., the result does not depend upon a concrete representation of $\tilde{x}$ as $\sigma \tilde{x}_0$, $\tilde{x}_0 \in \tilde{U}_i^*$. Let $\sigma \tilde{x}_1 = \tilde{x}$, $\tilde{x}_1 \in \tilde{U}_i^*$. Then $\tilde{x}_1 = \sigma_i^k \tilde{x}_0$ for some integer k and $\sigma_1 \sigma_i^k \tilde{x}_0 = \sigma \tilde{x}_0$, $\sigma_1 \sigma_i^k = \sigma$, $\sigma_1 = \sigma \sigma_i^{-k}$; now (3.2.2) implies that $f_i(\tilde{x}_1, \sigma_1) = f_i(\tilde{x}, \sigma)$.

Note that if $\tilde{x} \in p^{-1}(U_i^*)$, $\tilde{x} = \sigma \tilde{x}_0$, $\tilde{x}_0 \in \tilde{U}_i^*$, then

$$g_i(\tau \tilde{x}) = g_i(\tau \sigma \tilde{x}_0) = (\tilde{x}_0 - a_i)^{E_i} \chi \left((\tau \sigma)^{-1} \right) =$$
$$= (\tilde{x}_0 - a_i)^{E_i} \chi \left(\sigma^{-1} \right) \chi \left(\tau^{-1} \right) = g_i(\tilde{x}) \chi \left(\tau^{-1} \right),$$

thus (slightly changing the letters)

$$g_i(\sigma \tilde{x}) = g_i(\tilde{x}) \chi \left(\sigma^{-1} \right).$$

The next step is to define

$$\psi_i^E : p^{-1}(U_i^*) \times \mathbb{C}^p \to U_i^* \times \mathbb{C}^p,$$
$$\psi_i^P : p^{-1}(U_i^*) \times GL(p, \mathbb{C}) \to U_i^* \times GL(p, \mathbb{C})$$

as

$$\psi_i^E(\tilde{x}, y) = (x, g_i(\tilde{x})y), \quad \psi_i^P(\tilde{x}, Y) = (x, g_i(\tilde{x})Y),$$

where $x = p\tilde{x}$. If $(\tilde{x}, y) \sim (\tilde{x}_1, y_1)$ (in the same sense as in the Section 3.1), i.e., if $\tilde{x}_1 = \sigma \tilde{x}$, $y_1 = \chi(\sigma)y$ with some $\sigma \in \Delta$, then $\psi_i^E(\tilde{x}, y) = \psi_i^E(\tilde{x}_1, y_1)$:

$$\psi_i^E(\tilde{x}_1, y_1) = \psi_i^E(\sigma\tilde{x}, \chi(\sigma)y) = (x, g_i(\sigma\tilde{x})\chi(\sigma)y) =$$
$$= (x, g_i(\tilde{x})\chi\left(\sigma^{-1}\right)\chi(\sigma)y) = (x, g_i(\tilde{x})y) = \psi_i^E(\tilde{x}, y).$$

Analogously,

$$\psi_i^P(\sigma\tilde{x}, \chi(\sigma)Y) = \psi_i^P(\tilde{x}, Y),$$

i.e., if $(\tilde{x}, Y) \sim (\tilde{x}_1, Y_1)$ then $\psi_i^P(\tilde{x}, Y) = \psi_i^P(\tilde{x}_1, Y_1)$. In other words, ψ_i^E, ψ_i^P map the whole equivalence class into the same point. Thus we get maps (3.2.1).

These maps provide us with the corresponding glueings and so we get the bundles G, R with projections p_G, p_R, natural inclusions

$$i_E : E \to G, \quad i_P : P \to R$$

and identifications

$$p_G^{-1}(U_i) = U_i \times \mathbb{C}^p, \quad p_R^{-1}(U_i) = U_i \times GL(p, \mathbb{C}).$$

Having the latter in mind, we can say that the restrictions

$$i_E|p_E^{-1}(U_i^*) = \omega_i^E, \quad i_P|p_P^{-1}(U_i^*) = \omega_i^P.$$

Points

$$a = (x, y) \in U_i \times \mathbb{C}^p \subset G, \quad A = (x, Y) \in U_i \times GL(p, \mathbb{C}) \subset R$$

have local coordinates $\Phi_i^G(a)$, $\Phi_i^R(A)$ which are just the same points (x, y), (X, Y). But if we consider points

$$a \in p_E^{-1}(U_i^*), \, A \in p_P^{-1}(U_i^*)$$

as the points of $p_G^{-1}(U_i)$, $p_R^{-1}(U_i)$, then the local coordinates Φ_i^G, Φ_i^R of these points,– or, if you want, of the points $i_E a$, $i_P A$, – are $\omega_i^E(a)$, $\omega_i^P(A)$.

Now we are able to check that $Z(\tilde{x})$, i.e. $p_2\Phi_i^R(Z(\tilde{x}))$, *really has at most polynomial growth* when $x \to a_i$. Fix a sector $\Sigma \subset U_i^*$ with the vertex a. In terms of the polar coordinates (r, φ) with the origin at a_i,

$$\Sigma = \{x = a_i + re^{i\varphi}; \, 0 < r < \varepsilon, \, \varphi_1 \leq \varphi \leq \varphi_2\}.$$

Let $\tilde{\Sigma} \subset \tilde{S}$ be any sector covering Σ. It is contained in some connected component of $p^{-1}(U_i^*)$. Let this component be σU_i^*. Then $\sigma^{-1}\tilde{\Sigma}$ is a sector covering Σ and lying in $\tilde{U}_i^*$. Denote by $\tilde{x}$ the only point in $p^{-1}x \cap \sigma^{-1}\tilde{\Sigma}$ (we shall write $\tilde{x} = p^{-1}x \cap \Sigma$) and let $\tilde{x}_0 = \sigma^{-1}\tilde{x} = p^{-1}x \cap \sigma^{-1}\tilde{\Sigma}$. This $\tilde{x}_0$ is a single-valued function of $x \in \Sigma$. Denoting the unit element of Δ by ε_Δ, we have

$$g_i(\tilde{x}_0) = f_i(\tilde{x}_0, \varepsilon_\Delta) = (\tilde{x}_0 - a_i)^{E_i}\chi^{-1}(\varepsilon_\Delta) = (\tilde{x}_0 - a_i)^{E_i},$$
$$\Phi_i^R(i_P Z(\tilde{x})) = \omega_i^R(Z(\tilde{x})) = \omega_i^R(\langle(x, I)\rangle) = \omega_i^R(\langle(\sigma\tilde{x}_0, I)\rangle) =$$
$$= \omega_i^R(\langle(\tilde{x}_0, \chi(\sigma^{-1}))\rangle) = \psi_i^R(\tilde{x}_0, \chi(\sigma^{-1})) = (x, g_i(\tilde{x}_0)\chi(\sigma^{-1})) =$$
$$= (x, (\tilde{x}_0 - a_i)^{E_i}\chi(\sigma^{-1})),$$
$$p_2\Phi_i^R(i_P Z(\tilde{x})) = (\tilde{x}_0(x) - a_i)^{E_i}\chi(\sigma^{-1}) =$$
$$= \exp(E_i \ln(\tilde{x}_0(x) - a_i))\chi(\sigma^{-1}) = \exp(E_i \ln(re^{i\varphi}))\chi(\sigma^{-1}).$$

On $\sigma^{-1}\tilde{\Sigma}$

$$\ln re^{i\varphi} = \ln r + i\varphi + 2\pi ik$$

with some fixed k. Thus

$$p_2\Phi_i^R(i_P Z(\tilde{x})) = \exp(E_i \ln r)\exp(i\varphi E_i)G_i^k\chi(\sigma^{-1}).$$

$G_i^k\chi(\sigma^{-1})$ does not depend on x; the norm of $\exp(i\varphi E_i)$ is uniformly bounded as well as the norm of the inverse matrix, because $\varphi_1 \leq \varphi \leq \varphi_2$. The only factor which can be unbounded is $\exp(E_i \ln r) = r^{E_i}$. This is the matrix function which is well-known in linear algebra; we need not enter into details about it. Clearly its growth can be at most polynomial.

Now we shall discuss the situation with W. Essentially we need to know that there exists a meromorphic cross-section of the bundle $R \to \bar{\mathbb{C}}$. Indeed, multiplying such a cross-section by a suitable rational function, we can get a new meromorphic cross-section W such that all its singularities will be among $\{a_1, \ldots, a_n\}$. So $W|S$ is a "true" holomorphic cross-section of the bundle $P \to S$; thus we obtain a holomorphic map $Y : \tilde{S} \to GL(p, \mathbb{C})$ such that $Z(\tilde{x}) = W(p\tilde{x})Y(\tilde{x})$ for all $\tilde{x} \in \tilde{S}$. For $\tilde{x} \in p^{-1}(U_i^*)$

$$Y(\tilde{x}) = \left(p_2\Phi_i^R(i_P W(x))\right)^{-1}\left(p_2\Phi_i^R(i_P Z(\tilde{x}))\right) \quad (x = p\tilde{x}),$$

and as W is meromorphic, the first factor grows at most polynomially. We have already seen that the same is true for the second factor; hence it is true also for Y.

Let W be a holomorphic cross-section of R in a neighborhood of a_i, then $p_2 \Phi_i^R (i_P W(x))$ is holomorphically invertible at a_i. Since $p_2 \Phi_i^R (i_P Z(\tilde{x})) =$
$= (\tilde{x}_0(x) - a_i)^{E_i} \chi(\sigma^{-1}) = \chi(\sigma^{-1})(\tilde{x}_0(x) - a_i)^{E_i'}$ with $E_i' = \chi^{-1}(\sigma^{-1}) E_i \chi(\sigma^{-1})$, we can apply to $Y(\tilde{x})$ the statement of Corollary 2.2.1 (with $V(x) = (p_2 \Phi_i^R (i_P W(x)))^{-1} \chi(\sigma^{-1}), \Phi = I)$. Thus *if W is a holomorphic cross-section of R, then the corresponding system with the solution $Y(\tilde{x})$ is Fuchsian.*

But in general case the bundle R is not holomorphically trivial, therefore a global holomorphic cross-section W does not exist.

The existence of meromorphic cross-sections is proved in algebraic geometry in a much more general setting than the case which we need ($R \to \bar{\mathbb{C}}$). In our case it is a consequence of the Birkhoff-Grothendieck theorem. This theorem in its geometric form claims that *every complex holomorphic vector bundle G over $\bar{\mathbb{C}}$ is a direct (Whitney) sum of linear (i.e., one-dimensional) complex holomorphic vector bundles*

$$G = \mathcal{O}(j_1) \oplus \mathcal{O}(j_2) \oplus \cdots \oplus \mathcal{O}(j_p), \quad j_1 \geq \cdots \geq j_p, \qquad (3.2.3)$$

where for each integer j the linear bundle $\mathcal{O}(j)$ can be obtained as follows. Take direct products

$$\mathbb{C} \times \mathbb{C}, \quad (\bar{\mathbb{C}} \setminus \{0\}) \times \mathbb{C}$$

and glue them over $\bar{\mathbb{C}} \setminus \{0\}$ using the following equivalence: point $(x, y) \in \mathbb{C} \times \mathbb{C}$ is equivalent to the point

$$(x, x^j y) \in (\bar{\mathbb{C}} \setminus \{0\}) \times \mathbb{C}.$$

It follows that G has the following transformation function $g_{\infty,0}$ describing the change of some coordinates corresponding to the coordinate neighborhoods $V_0 = \mathbb{C}$, $V_\infty = \bar{\mathbb{C}} \setminus \{0\}$:

$$g_{\infty,0}(x) = \begin{pmatrix} x^{j_1} & & & 0 \\ & \cdots & & \\ & & \cdots & \\ 0 & & & x^{j_p} \end{pmatrix}$$

Clearly this implies the existence of meromorphic cross-sections to R. Take a cross-section over V_0 which in terms of the corresponding local coordinates is $x \mapsto (x, I)$. Then its coordinate expression over V_∞ is given by the same matrix $g_{\infty,0}(x)$ which is the rational function of x (and of $z = 1/x$ which is the natural coordinate in V_∞).

Moreover obviously this cross-section is holomorphic outside the point ∞. If one wants to obtain a holomorphic outside a point a_i cross-section he must start from the cross-section $x \rightarrow (x, G_i)$, where

$$
G_i(x) = \begin{pmatrix} (x - a_i)^{-j_1} & & 0 \\ & \ddots & \\ 0 & & (x - a_i)^{-j_p} \end{pmatrix}
$$

Obviously this cross-section is holomorphic at ∞, since its coordinate expression over V_∞ is given by the matrix $g_{\infty,0}(x)G_i = (x, I + O(1))$ at ∞. So due to the above statement we have the following result, which for the first time also was obtained by Plemelj:

Theorem 3.2.1 *For each a_i there always exists a regular system (1.2.1) with given singular points $a_1, \ldots, a_n$ and given monodromy, which is Fuchsian outside of a_i.*

The last step of the proof of Plemelj's result (see (1) in Section 1.2) is given in [ArII]. Here it is. Let the matrix G_i have a diagonal form in the basis of columns of $Y(\tilde{x})$, $\tilde{x} \in \tilde{U}_i^*$. Consider factorization (2.2.3) for the matrix $Y(\tilde{x})$:

$$
Y(\tilde{x}) = Z_i(x)(\tilde{x} - a_i)^{E_i}.
$$

By Sauvage's lemma there exists a holomorphically invertible outside of a_i matrix $\Gamma(x)$ such that

$$
\Gamma(x)Z_i(x) = V_i(x)(x - a_i)^{\Phi},
$$

where $V_i(x)$ is holomorphically invertible at a_i, $\Phi = \mathrm{diag}(\varphi^i)$. (One can easily prove this lemma combining Birkhoff-Grothendieck's theorem with our Lemma 4.1.3. There is also an elementary direct proof avoiding use of this theorem; see, e.g., [Ha]). Introduce a new dependent variable $t = \Gamma y$. Evidently,

$$
\frac{dt}{dx} = \left(\frac{d\Gamma}{dx} \cdot \Gamma^{-1} + \Gamma A(x)\Gamma^{-1} \right) t. \tag{3.2.4}
$$

Therefore the new system is still Fuchsian outside of a_i. Since $L = \Phi + x^\Phi E_i \cdot x^{-\Phi} = \Phi + E_i$ (E_i is a diagonal matrix), L is holomorphic and from Corollary 2.2.1 we obtain that this system is Fuchsian at the point a_i too. This completes the proof of Plemelj's result.

The algebraic-geometrical proof of the Birkhoff-Grothendieck theorem begins with the proof of the existence of meromorphic cross-sections to the vector bundle $G \to \bar{\mathbb{C}}$ (this is perhaps the most important part of the proof) and involves the analysis of the properties of such cross-sections. Although this is not the same as the existence of meromorphic cross-sections to R, these things seem to be quite close to each other; and although it is only the existence of a meromorphic cross-section for the bundle R which we need, this is perhaps only slightly weaker than the Birkhoff-Grothendieck theorem. Beside this, the latter provides some additional information which may be useful, although until now there was not much use for this information.

We shall mention one case in which the additional information provided by the Birkhoff-Grothendieck theorem plays some role. This is the case $p = 2$. Applying to the monodromy $\chi : \Delta \to GL(p, \mathbb{C})$ the construction of Sections 3.1, 3.2, we get some bundle G. According to the Birkhoff-Grothendieck theorem

$$G = \mathcal{O}(j_1) \oplus \mathcal{O}(j_2)$$

with some j_1, j_2. The number $j_1 + j_2$ is the well-known topological invariant of the bundle G (Chern number). Having the sum, it is naturally to pay attention to the difference. In [Bo2] it was found out that $|j_1 - j_2|$ coincides with the "Fuchsian weight" γ_χ. This statement will be proved in Chapter 6.

3.3 Proof of the Birkhoff-Grothendieck theorem

This theorem has two forms – geometric (which is due to Grothendieck and which was already formulated in Section 3.2) and analytic (which is due to Birkhoff). In order to formulate the latter let us introduce the following notation:

$$K = K(r, R) := \{x; r \le |x| \le R\},$$
$$D = D(R) := \{x; |x| \le R\}, \quad C = C(r) := \{x; |x| \ge r\} \cup \{\infty\},$$
$$H(K), H(D), H(C) := \{\text{continuous functions } K \to \mathbb{C}, \ D \to \mathbb{C},$$
$$C \to \mathbb{C} \text{ which are holomorphic in Int } K, \text{ Int } D, \text{ Int } C\}$$

(Int means the interior),

$$V(K), V(D), V(C) \quad := \quad \{\text{vector functions with the same properties,}$$
$$\text{i.e., whose components lie in}$$
$$H(K), H(D), H(C)\},$$
$$M(K), M(D), M(C) \quad := \quad \text{the corresponding } (p,p)\text{-matrix functions,}$$
$$MI(K), MI(D), MI(C) \quad := \quad \{\text{matrix functions from } M(K), M(D), M(C)$$
$$\text{which are invertible (everywhere in} K, D, C)\},$$
$$H_0(C), M_0(C) \quad := \quad \{\text{functions } f \in H(C), M(C) \text{ such that}$$
$$f(\infty) = 0\}.$$

Note that H, V, M, H_0, M_0 are the Banach spaces with the usual C-norm.

The analytic version claims: *Let $A \in MI(K)$. Then there exists $U \in MI(C)$, $W \in MI(D)$ such that everywhere in K*

$$A = U \begin{pmatrix} x^{j_1} & & & & 0 \\ & \ddots & & & \\ & & \ddots & & \\ & & & \ddots & \\ 0 & & & & x^{j_p} \end{pmatrix} W = U x^J W, \ J := diag(j_i), \ j_1 \geq \cdots \geq j_p$$

$$(3.3.1)$$

with some integers $j_1, \ldots, j_p$.

The analytic version is equivalent to the geometric one – both claim that after a suitable change of the local coordinates over D and C the transformation function of arbitrary holomorphic vector bundle $G \to \bar{\mathbb{C}}$ becomes the transformation function for the bundle (3.2.3).

Note that the standard procedure of expanding the function from $H(K)$ into the sum of the Taylor and Laurent series provides a decomposition $f = g + h$ with $g \in H(D)$, $h \in H_0(C)$ and the corresponding projections $P : H(K) \to H(D)$, $Q : H(K) \to H_0(C)$ are bounded linear operators. (Former P, Q will not be used any more, as well as former E, F). Perhaps it is worth while to explain why g has the required continuity properties near the circle $|x| = R$ and h– near the circle $|x| = r$, as the construction of g, h involves some integrals along the circle $|x| = \rho$, $r < \rho < R$, and the corresponding estimates do not "work" up to the circles bounding K. But this construction implies the continuity (and even analyticity) of h near the circle $|x| = R$ and of g near the circle $|x| = r$. It remains to use the identities

$$g = f - h, \quad h = f - g$$

and the continuity of f. A slightly more detailed elaboration of the same idea proves the boundedness of P, Q.

Evidently we have an analogous decomposition for elements of $M(K)$. The corresponding projection $M(K) \rightarrow M_0(C)$ also shall be denoted by Q.

Let us prove our theorem for $A \in M(K)$ having the form $A = I + B$ with sufficiently small B ("small" in sense of the norm in $M(K)$). (Note that such A is always invertible). We shall see that in this case in (3.3.1) there will be no diagonal matrix x^J (i.e., all $j_i = 0$).

It is sufficient to prove the existence of $X \in M(C)$ such that it is small (in $M_0(C)$) and

$$Q(A(I + X)) = 0. \tag{3.3.2}$$

Indeed, this means that $U := A(I + X) \in M(D)$. Clearly this matrix is invertible (as A and $I + X$ are – the latter is invertible because X is small), so we have $A = UW$, where $W := (I + X)^{-1} \in MI(C)$.

In (3.3.2) Q acts on $I + B + X + BX$. But $QI = 0, QX = X$ (we are looking for X in $M_0(C)$). So we have to find X such that

$$X + QB + QBX = 0. \tag{3.3.3}$$

Denote the operator

$$M_0(C) \rightarrow M_0(C), \quad X \mapsto QBX$$

by T. This operator is small (in sense of the usual norm), as B is small and Q is bounded. Hence operator $E + T$, where E is the identity in $M_0(C)$, is invertible. But (3.3.3) means that $(E + T)X = -QB$.

An analogous argument proves that in the same case we have also $A = WU$ with some (new) $W \in MI(C), U \in MI(D)$. Instead of referring to the "same argument" one can apply the statement just proved to A^{-1} (which is clearly also close to I). If $A^{-1} = UW$ with $U \in MI(D), W \in MI(C)$, then $A = W^{-1}U^{-1} \in MI(D), W^{-1} \in MI(C)$.

Next statement is that any $A \in MI(K)$ can be represented in the form

$$A = UFV \tag{3.3.4}$$

where $U \in MI(D)$, $W \in MI(C)$ and F is a rational matrix function of x which is invertible in K. This means that a holomorphic vector bundle over $\bar{\mathbb{C}}$ is holomorphically equivalent to the bundle having purely algebraic description. Of course there exist much more general theorems asserting the equivalence of analytic and algebraic objects. But our statement admits an easy and elementary proof which, by the way, gives some additional information: there exists a representation (3.3.4) having the properties mentioned above and such that all poles of F lie in $(\bar{\mathbb{C}} \setminus D) \cup \{0\}$. We could do well without this information, but as we get it "at no extra cost", we shall use it.

We shall use the following fact: any function $f \in H(K)$ can be approximated uniformly in K by a rational function having poles only in $0, \infty$. This would be quite clear if $f \in H(r_1, R_1)$ with $r_1 < r$, $R_1 > R$, – then we can simply truncate the corresponding Taylor and Laurent series and this provides an uniform approximation in K. For $f \in H(K)$ let us write $f = g + h$, $g \in H(D)$, $h \in H_0(C)$. Consider $u(x) = g(x/(1 + \varepsilon))$, $v(x) = h((1 + \varepsilon)x)$. For sufficiently small ε both $g - u$ and $h - v$ can be made arbitrarily small in K. And now we can approximate u by a polynomial in $1/x$ and v by a polynomial in $1/x$, these approximations being uniform in D, resp. C.

The same applies to the matrix functions from $M(K)$. Now let $A \in MI(K)$ and B be a rational matrix function of x approximating A. Taking a sufficiently close approximation we can make not only $A - B$, but also $AB^{-1} - I$ and $B^{-1}A - I$ as small in K as we want.

For some B this difference becomes so small that we can apply our previous result. We get $B^{-1}A = W_1 U_1$ with some $W_1 \in MI(C)$, $U_1 \in MI(D)$. Thus $A = BW_1 U_1$. Now apply the same argument to BW_1 (which clearly belongs to $MI(K)$). For some rational H the difference $BW_1 H^{-1} - I$ is small whereas $H \in MI(K)$ and the only poles of H can be $0, \infty$. Then $BW_1 H^{-1}$ can be represented as

$$BW_1 H^{-1} = W_2 U_2, \quad W_2 \in MI(C), \quad U_2 \in MI(D).$$

Consequently,

$$U_2 = W_2^{-1} BW_1 H^{-1}, \tag{3.3.5}$$

$$A = W_2 U_2 HU_1 = W_2 FU_1, \quad F := U_2 H. \tag{3.3.6}$$

Not only does $U_2 \in MI(D)$, but U is also a rational matrix function with all its poles outside of D. Indeed, in (3.3.5) $W_2^{-1} \in MI(C)$, $W_1 \in MI(C)$, B and H are rational, so U_2 is meromorphic in Int$C$. But at the same time $U_2 \in MI(D)$. Thus U_2 is meromorphic on the whole $\mathbb{C}$, i.e., rational. Now in (3.3.6) $F = U_2 H$ is rational and of course $U_1 H \in MI(K)$, as $U_2 H = W_2^{-1} A U_1$, where all three factors belong to $MI(K)$. But we know also that $U_2 \in MI(D)$ and the only pole of H lying in D can be 0. This proves our additional remark about poles of F.

All these steps were of a preliminary character. They guarantee that it is sufficient to prove the theorem for the rational matrix function $A = F \in MI(K)$ and even for the case when all poles of F lie in $(\mathbb{C} \setminus D) \cup \{0\}$. Indeed, if in K our original $A = W_1 F U_1$ and $F = W_2 x^J U_2$ with $J =$diag(j_i), j_i integers, then

$$A = (W_1 W_2) x^J (U_2 U_1)$$

is the required representation for A.

Now the essential part of the proof begins. We want to find U, W such that $F = W x^J U$ in K (plus the usual conditions about U, W). In other words,

$$FU^{-1} = W x^J.$$

$$. \quad (3.3.7)$$

If U^{-1} and W have columns $d_i(x)$, respectively $c_i(x)$:
$U^{-1}(x) = (d_1(x), \ldots, d_p(x))$, $\quad W(x) = (c_1(x), \ldots, c_p(x))$, then (3.3.7) means that for each i

$$F(x)c_i(x) = x^{j_i} d_i(x) \text{ for all } x \in K.$$

This makes it reasonable to consider triples (j, c, d) such that j is an integer, $c \in V(C)$, $d \in V(D)$ and

$$F(x)c(x) = x^j d(x) \text{ for all } x \in K. \tag{3.3.8}$$

Besides this, the columns $c_i(x)$ of the invertible matrix $W(x)$ must be all $\neq 0$. This makes it reasonable to impose the additional requirement $c(\infty) \neq 0$.

This justifies the following definition. Let $y \in \mathbb{C}^p \setminus \{0\}$. *An admissible triple* for y is a triple (j, c, d) such that j is an integer, $c \in V(C)$, $d \in V(D)$, $c(\infty) = y$ and (3.3.8) holds. In such a triple not only $c \in V(C)$, but also the formula

$$c(x) = F^{-1}(x)x^j d(x)$$

provides a meromorphic extension of c in $\bar{\mathbb{C}} \setminus C$. Thus c is rational in x. For the same reason d is also rational in x.

Let us check that for any $y \in \mathbb{C}^p \setminus \{0\}$ there exists an admissible triple. Write F as

$$F(x) = P(x)/x^\beta q(x)$$

where β is a nonnegative integer, P is a matrix-valued polynomial in x (let its degree be α) and q is a usual (scalar) polynomial in x with the leading coefficient 1 and with the zeroes which are nonzero poles of F (with the same multiplicity). All zeroes of q lie outside of D. Thus

$$d(x) := \frac{1}{q(x)} P(x) y$$

is a rational vector function of x having no poles in D, and $(-\beta, y, d)$ is an admissible triple for y.

For any triple (j, c, d) which is admissible for some y

$$j \leq \alpha - \beta \qquad\qquad (3.3.9)$$

Indeed, rewrite (3.3.8) as

$$P(x)c(x) = x^{\beta+j}q(x)d(x).$$

Here the left hand side has no pole in $\mathbb{C} \setminus D$. Hence $q(x)d(x)$ has no pole there. But neither $d(x)$ nor $q(x)$ has poles in D. Thus $q(x)d(x)$ has no pole in $\mathbb{C}$, i.e., this is a polynomial $g(x)$, say, of degree l. When $x \to \infty$

$$x^\alpha(P_\alpha + o(1))(y + o(1)) = x^{\beta+j+l}(g_l + o(1)),$$

where P_α, g_l are leading terms of P, g. It follows that $\beta + j + l - \alpha \leq 0$ and this implies (3.3.9).

Let us call (j, c, d) *a maximal admissible triple for* $y = \mathbb{C}^p \setminus \{0\}$ if it is an admissible triple for y and if for any other admissible triple (i, a, b) for y one has $i \leq j$. Any $y \in \mathbb{C}^p \setminus \{0\}$ has maximal admissible triples. This follows from the fact that j's appearing in the admissible triples are bounded from above according to (3.3.9).

If (j, c, d) is a maximal admissible triple for y, then c has no zeroes in C and d has no zeroes in D. In order to prove this consider three cases.

a). $d(a) = 0$, $a \in K$, or $c(a) = 0$, $a \in K$. As F is invertible in K, it follows from (3.3.7) that in this case both c and d have zero at a. So

$$c(a) = (x - a)f(x), \quad d(a) = (x - a)h(x),$$

where $f \in V(C)$, $g \in V(D)$. Also $g := xf = xc/(x - a) \in V(C)$ and $g(x) \to y$, when $x \to \infty$. But

$$Fg = Fxc/(x - a) = x^{j+1}d/(x - a) = x^{j+1}h.$$

We see that $(j + 1, g, h)$ is an admissible triple for y, although $j + 1 > j$.

b). $d(a) = 0$, $a \in D \setminus K = \mathrm{Int}D(r)$. Then $d = (x - a)g$, $g \in V(D)$, and if $f := xc/(x - a)$, then $f \in V(C)$ and $f(\infty) = y$. But $Ff = x^{j+1}g$, so $(j + 1, f, g)$ is an admissible triple for y. This contradicts to the maximality of (j, c, d).

c). $c(a) = 0$, $a \in C \setminus D$ ($a \neq \infty$ because $c(\infty) = y \neq 0$). Then $c = (x - a)f$, $f \in V(C)$ and if $g := xf/(x - a)$, $h := g/(x - a)$, then $g \in V(C)$, $h \in V(D)$, $Fg = x^{j+1}h$, which contradicts to the maximality of (j, c, d).

Define $\varphi : \mathbb{C}^p \to \mathbb{Z}$ as following: $\varphi(0) = \infty$, and if $y \neq 0$, then $\varphi(y)$ is the integer j appearing in the maximal admissible triple (j, c, d) for y. *This φ has the same properties (2.2.8) as Levelt's φ used in Chapter 2.* We need only to check the property

$$\varphi(y_1 + y_2) \geq \min(\varphi(y_1), \varphi(y_2)). \tag{3.3.10}$$

Properties occurring in the first line of (2.2.8) are evident, and the last property in the second line (referring to the case $\varphi(y_1) \neq \varphi(y_2)$) follows from (3.3.10) and $\varphi(y) = \varphi(-y)$. Indeed, let (3.3.10) be already proven and let $\varphi(y_1) < \varphi(y_2)$. Then $\min(\varphi(y_1), \varphi(y_2)) = \varphi(y_1)$. So $\varphi(y_1) \leq \varphi(y_1 + y_2)$. But

$$\varphi(y_1) = \varphi((y_1 + y_2) - y_2) \geq \min(\varphi(y_1 + y_2), \varphi(y_2)),$$

and the latter min must be $\varphi(y_1 + y_2)$, as it cannot happen that $\varphi(y_1) \geq \varphi(y_2)$. So $\varphi(y_1) \leq \varphi(y_1 + y_2) \leq \varphi(y_1)$, which implies $\varphi(y_1) = \varphi(y_1 + y_2)$.

Let us prove (3.3.10). In the case when one of the vectors $y_1, y_2, y_1 + y_2$ is 0 it is trivial, so we may assume that all they are $\neq 0$. Let $\varphi(y_1) = j \leq \varphi(y_2) = k$ and let (j, c, d), (k, f, g) be the maximal admissible triples for y_1, y_2. Then $c + f \in V(C)$, $(c + f)(\infty) = y_1 + y_2$, $d + x^{k-j}g \in V(D)$, and in K

$$F(c + f) = x^j d + x^k g = x^j \left(d + x^{k-j} g\right).$$

Thus $\varphi(y_1 + y_2) \geq j = \min(j, k)$.

As in Section 2.2, there exists a filtration

$$0 = E_0 \subset E_1 \subset \cdots \subset E_h = \mathbb{C}^p$$

such that φ is constant on $E_i \setminus E_{i-1}$ and if $\psi^j = \varphi(E_j \setminus E_{j-1})$ then $\psi^0 := \infty > \psi^1 > \cdots > \psi^h$. We say again that φ takes the value ψ^j with the multiplicity $k_j := \dim E_j - \dim E_{j-1}$, and introduce the same φ^i as in Section 2.2 – each φ^i is some ψ^j, there are k_j φ^i's equal to ψ^j and $\varphi^1 \geq \cdots \geq \varphi^p$. There exists a basis $y_1, \ldots, y_p$ in $\mathbb{C}^p$ such that $\varphi(y_i) = \varphi^i$.

Let (φ^i, c_i, d_i) be the maximal admissible triple for y_i. It turns out that for any $x \in \mathbb{C}$ vectors $c_i(x)$ are linearly independent and for any $x \in D$ vectors $d_i(x)$ are linearly independent. This follows from the fact that any nontrivial linear combination of c_i is a c appearing in some maximal admissible triple, and any nontrivial linear combination of d_i is a d appearing in some maximal admissible triple. Here is the proof of this fact. Consider

$$c(x) := \sum \{\lambda_i c_i(x); i \in I\},$$

where $I \subset \{1, \ldots, p\}$ is nonempty and $\lambda_i \neq 0$ for all $i \in I$. Denote

$$\psi := \min\{\varphi^i; i \in I\}$$

(this ψ equals to one of our ψ^j, say, to ψ^s),

$$J := \{i \in I; \varphi^i = \psi\} \quad (J \text{ is nonempty}),$$
$$d(x) := \sum \{\lambda_i x^{\varphi_i - \psi} d_i(x); i \in I\},$$
$$y := c(\infty) = \sum\{\lambda_i c_i(\infty); i \in I\} = \sum\{\lambda_i y_i; i \in I\}.$$

Then $c \in V(C)$, $d \in V(D)$ (because $\varphi^i - \psi \geq 0$ for all $i \in I$),

$$Fc = x^\psi Fd,$$

$y \neq 0$ (because y_i is a basis and J is nonempty) and (ψ, c, d) is an admissible triple for y. It is even the maximal admissible triple for y, which means that $\varphi(y) = \psi$. Indeed,

$$\varphi(y) \geq \min\{\varphi(y_i); i \in I\} = \psi,$$

but if $\varphi(y) > \psi = \psi^s$, then $y \in E_{s-1}$, although our choice of the basis $y_1, \ldots, y_p$ is such that $\{y_i : i \in J\}$ are linearly independent modulo E_{s-1}.

Finally consider

$$d(x) := \sum \{\lambda_i d_i(x); i \in I\},$$

where $I \subset \{1, \ldots, p\}$ is nonempty and $\lambda_i \neq 0$ for all $i \in I$. Denote

$$\psi := \min\{\varphi^i; i \in I\}$$

(this ψ is equal to one of ψ^j, say, to ψ^s),

$$J := \{i \in I; \varphi^i = \psi\} \quad (J \text{ is nonempty}),$$
$$c(x) := \sum \{\lambda_i x^{\psi - \varphi_i} c_i(x); i \in I\}.$$

Then $c \in V(C)$ (because $\psi - \varphi^i \leq 0$ for all $i \in I$), $d \in V(D)$,

$$Fc = x^\psi d.$$

Also $y := c(\infty)$ makes sense and $y := \sum \{\lambda_i y_i; i \in J\} \neq 0$, (because y_i is a basis and J is nonempty). So (ψ, c, d) is an admissible triple for y. It is even the maximal admissible triple for y, which means that $\varphi(y) = \psi$. Indeed

$$\varphi(y) \geq \min\{\varphi(y_i); i \in I\} = \psi,$$

but if $\varphi(y) > \psi = \psi^s$, then $y \in E_{s-1}$, although our choice of the basis $y_1, \ldots, y_p$ is such that $\{y_i; i \in J\}$ are linearly independent modulo E_{s-1}.

After this we can take $U^{-1} := (d_1, \ldots, d_p)$, $W := (c_1, \ldots, c_p)$, and this provides (3.3.7) with all the required properties of U, W.

The complete version of the Birkhoff-Grothendieck theorem contains a uniqueness condition for the numbers $j_1, \ldots, j_p$. This means that the following statement holds: *Let a matrix $A \in MI(K)$ have two different decompositions*

$$A = U_i x^{J_i} W_i, \quad i = 1, 2 \tag{3.3.11}$$

of the form (3.3.1). Then $J_1 = J_2$.

Denote by $j_1^i, \ldots, j_p^i$ coefficients of the corresponding matrix J_i. It follows from (3.3.11) that

$$U_2^{-1} U_1 = x^{J_2} W_2 W_1^{-1} x^{-J_1}, \tag{3.3.12}$$

therefore the matrices $U_2^{-1} U_1$ and $W_2 W_1^{-1}$ can be analytically continued in $\bar{\mathbb{C}} \setminus \{\{0\} \cup \{\infty\}\}$ and their continuations (which we denoted by the same symbols) are holomorphically invertible there. So we have that coefficients (u_{km}) of $U_2^{-1} U_1$ are holomorphic outside of 0 and may be presented at $x = 0$ as follows:

$$u_{km}(x) = x^{j_k^2 - j_m^1} w_{km}(x), \tag{3.3.13}$$

where (w_{km}) are coefficients of $W_2 W_1^{-1}$.

Assume that $J_2 \neq J_1$. Without loss of generality we may assume that $j_k^2 = j_k^1$ for $k < l$ and $j_l^2 > j_l^1$ for some $l \leq p$. Since the functions $w_{km}(x)$ are holomorphic at 0, from (3.3.13) in this case we obtain $u_{km}(0) = 0$ for $k = 1, \ldots, l; m = l, \ldots, p$. Since $u_{km}(x)$ is holomorphic outside of $x = 0$ we conclude that $u_{km}(x) \equiv 0$ for $k = 1, \ldots, l; m = l, \ldots, p$. Therefore the matrix $U_2^{-1} U_1$ has the form

$$U_2^{-1} U_1 = \begin{pmatrix} * & & 0 & \cdots & 0 \\ & \ddots & \vdots & & \vdots \\ & & 0 & \cdots & 0 \\ & & & \ddots & \\ * & & & & * \end{pmatrix} \begin{matrix} \\ \\ -l \\ \\ \end{matrix}$$

$$\begin{matrix} | \\ l \end{matrix}$$

and we have $\det U_2^{-1} U_1(x) \equiv 0$ (since its rows with numbers $1, \ldots, l$ are linearly dependent) that contradicts the condition $U_2^{-1} U_1 \in MI(C)$. This contradiction means that our assumption is false, i.e., $J_1 = J_2$.

3.4 Some other known results

Using Levelt's factorization (2.2.25) we can easily prove that for $p = 2$ Hilbert's 21st problem has a positive solution independently on n. (Recall that this result follows from [Dek], but Dekkers did not use Levelt's factorization, so from our point of view his proof was more complicated, than the proof we are presenting here).

Indeed, let $p = 2$ and let $Y(x)$ be a fundamental matrix to some regular system (1.2.1) with the given monodromy χ. By Plemelj's result we may assume that the system is Fuchsian at $a_2, \ldots, a_n$. Due to Plemelj's theorem it is sufficient to prove the statement only for nondiagonalizable monodromy matrices $\chi(\sigma_1), \ldots, \chi(\sigma_n)$.

Consider factorizations (2.2.25) for all $i = 1, 2, \ldots, n$. By Theorem 2.2.1 the matrices $V_2(a_2), \ldots, V_n(a_n)$ are nondegenerate. Let the matrix $V_1(x)$ have the form

$$V_1(x) = \begin{pmatrix} 1 & v_{12} \\ c_1(x-a_1)^{k_1} & c_2(x-a_1)^{l_1} \end{pmatrix}(1+O(1)), \quad k_1 > 0, \ l_1 \geq 0. \quad (3.4.1)$$

(The first column of the matrix $V_1(a_1)$ is always nonzero (see Section 2.2), so by transformation (3.2.4) with a suitable $\Gamma = S$ we can obtain the matrix $\tilde{Y} = SY$ with $\tilde{V}_1 = SV_1$ of the form (3.4.1)).

By v_{kl}^i denote elements of the matrices $V_i(x)$ and by s denote the sum of exponents of the system over all $a_1, \ldots, a_n$. From Theorem 2.2.2 we have

$$s \leq 0. \tag{3.4.2}$$

Let $c_1 \neq 0$, $l_1 > 0$ in (3.4.1). Transform our system to the system with the fundamental matrix $Y' = \Gamma Y$, where

$$\Gamma_1 = \begin{pmatrix} 1 & -\frac{1}{c_1(x-a_1)^{k_1}} \\ 0 & 1 \end{pmatrix}.$$

The matrix Γ_1 is holomorphically invertible off a_1, therefore Y' has a factorization (2.2.25) with $V_i' = \Gamma \cdot V_i$ and with the same Φ_i for $i = 2, \ldots, n$. At a_1 we get

$$\Gamma V_1(x) = \begin{pmatrix} c_3(x-a_1)^t & -\frac{c_2}{c_1}(x-a_1)^{l_1-k_1} + v_{12} \\ c_1(x-a_1)^{k_1} & c_2(x-a_1)^{l_1} \end{pmatrix}(1+O(1)), \quad t > 0.$$

If $t < k_1$ transform Y' to $\tilde{Y} = \Gamma_2 Y'$ by

$$\Gamma_2 = \begin{pmatrix} 1 & -\frac{c_3}{c_1(x-a_1)^{k_1-t}} \\ 0 & 1 \end{pmatrix}$$

and so on.

After finite steps of such transformations we obtain the system with a fundamental matrix $Y^1(x)$ such that the following conditions hold:

i) the system is Fuchsian at $a_2, \ldots, a_n$ with the same exponents at these points,

ii) the matrix V_1^1 from factorization (2.2.25) has the form:

$$V_1^1(x) = \begin{pmatrix} c(x-a_1)^{k_1} & -\frac{c_2}{c_1}(x-a_1)^{l_1-k_1} + v_{12} \\ c_1(x-a_1)^{k_1} & c_2(x-a_1)^{l_1} \end{pmatrix} (1 + O(1)). \qquad (3.4.3)$$

If $l_1 < k_1$, then

$$V_1^1 = (\text{hol.})(x-a_1)^{\begin{pmatrix} k_1 & 0 \\ 0 & l_1 - k_1 \end{pmatrix}}, \qquad k_1 > l_1 - k_1.$$

If $l_1 \geq k_1$, then

$$V_1^1 = (\text{hol.})(x-a_1)^{\begin{pmatrix} k_1 & 0 \\ 0 & 0 \end{pmatrix}}, \qquad k_1 > 0.$$

In both cases from (2.2.25) we get

$$Y_1^1 = (\text{hol.})(x-a_1)^{\Phi_1'}(\tilde{x}_0 - a_1)^{E_1} \chi(\sigma^{-1}),$$

where Φ_1' is the matrix of valuations for the new system. But

$$tr\,\Phi_1' \geq tr\,\Phi_1 + k_1 + (l_1 - k_1) = tr\,\Phi_1 + l_1 > tr\,\Phi_1$$

or $(tr\,\Phi_1' \geq tr\,\Phi_1 + k_1 > tr\,\Phi_1)$. Thus due to i) we obtain that the sum s' of exponents of the new system is greater than s:

$$s < s' \le 0. \tag{3.4.4}$$

If $s' = 0$, then by Theorem 2.2.2 this system is already Fuchsian at all points. If $s' < 0$ and $c_1 \ne 0$ in formula (3.4.1) for this system, then we may again apply the same procedure and so on until we get $\tilde{s} = 0$ or $\tilde{c}_1 = 0$ in some step.

Let in (3.4.2) $c_1 = 0$. Then by (2.2.25)

$$Y_1(\tilde{x}) = \begin{pmatrix} y_{11} & y_{12} \\ 0 & y_{22} \end{pmatrix} \tag{3.4.5}$$

in $\tilde{U}_1^*$. By the theorem on uniqueness for analytic functions we have $y_{21} \equiv 0$ for all $\tilde{x} \in \tilde{S}$. But this means that the vector $\begin{pmatrix} y_{11} \\ 0 \end{pmatrix}$ is a common eigenvector for all σ_j^*. Indeed,

$$\sigma_j^* \begin{pmatrix} y_{11} \\ 0 \end{pmatrix} = \lambda_1 \begin{pmatrix} y_{11} \\ 0 \end{pmatrix} + \lambda_2 \begin{pmatrix} y_{12} \\ y_{22} \end{pmatrix},$$

but of course $0 \circ \sigma_j = 0$, therefore $\lambda_2 = 0$ (in opposite case we would have the impossible identity $y_{22} \equiv 0$, which contradicts the inequality $\det Y(\tilde{x}) \ne 0$). Thus in the case $c_1 = 0$ the representation χ is reducible. Therefore we can simultaneously transform $\chi(\sigma_i)$ to the form

$$\chi(\sigma) = \begin{pmatrix} \lambda_1^i & * \\ 0 & \lambda_2^i \end{pmatrix}, \quad i = 1, \ldots, n.$$

Since by assumption all $\chi(\sigma_i)$ are nondiagonalizable, we get $\lambda_1^i = \lambda_2^i$ for all i. Thus, $[\chi(\sigma_i), \chi(\sigma_j)] = 0$ and we obtain that χ is the commutative representation.

Consider the matrix

$$Y(\tilde{x}) = (x - a_1)^{-\mu} \prod_{i=1}^{n} (x - a_i)^{E_i}, \tag{3.4.6}$$

where E_i is from (2.2.2), $\mu = \sum_{i=1}^{n} \mu_i$. The system with this fundamental matrix has the form

$$\frac{dy}{dx} = \left[\left(\frac{E_1 - \mu I}{x - a_1} \right) + \sum_{j=2}^{n} \frac{E_j}{x - a_j} \right] y \tag{3.4.7}$$

and it satisfies equality (1.2.3).

Indeed, we have $\prod_{i=1}^{n} \chi(\sigma_i) = I$, therefore

$$\frac{1}{2\pi i} \cdot \ln \left(\prod_{i=1}^{n} \chi(\sigma_i) \right) = \sum_{i=1}^{n} E_i = \mu I, \quad \mu = \sum_{i=1}^{n} \mu_i \in \mathbb{Z}.$$

Thus we get a Fuchsian system with the given χ too.

Remark 3.4.1 *Formula (3.4.6) provides a positive answer on Hilbert's 21st problem for a commutative representation χ independently of p.*

4 Irreducible representations

4.1 Technical preface

In this subsection we present some technical statements which will be used below.

The way of the solution of Hilbert's 21st problem for an irreducible representation consists in improving of a regular system with help of suitable transformations of a depending variable

$$z = \Gamma(x)y .$$ (4.1.1)

Under this transformation system (1.2.1) is transformed to the following one:

$$\frac{dz}{dx} = A'(x)z,$$

where

$$A'(x) = \frac{d\Gamma}{dx}\Gamma^{-1} + \Gamma A \Gamma^{-1}.$$ (4.1.2)

If the matrix Γ is holomorphically invertible at some point b, then the matrix $A'(x)$ has the same type of singularity at b as $A(x)$. (If A is holomorphic at b, so does A'). Taking $\Gamma(x)$ to be holomorphically invertible off a_i , we can try to improve the type of singularity of (1.2.1) at a_i without changing for the worse at other points. For this purpose we need the following statements.

Lemma 4.1.1 *Suppose that the matrix $W(x)$ of the size $(p-l, l)$ is holomorphic, and the matrix $Y(x)$ of the size (l, l) is holomorphically invertible in the neighborhood U_i of the point a_i. For any integer-valued diagonal matrix $C = diag(c_1, \ldots, c_p)$, there exists a matrix-function $\Gamma(x)$, meromorphic on $\bar{\mathbb{C}}$ and holomorphically invertible off the point a_i, such that*

$$\Gamma(x)(x-a_i)^C \left(\begin{array}{c} Y(x) \\ W(x) \end{array} \right) = (x-a_i)^{C'} \left(\begin{array}{c} Y(x) \\ W'(x) \end{array} \right),$$ (4.1.3)

where $C' = diag(c_1, \ldots, c_l, c'_{l+1}, \ldots, c'_p)$, $c'_j > \min(c_1, \ldots, c_l)$, $j = l+1, \ldots, p$, $W'(x)$ is a matrix holomorphic in U_i.

Proof. We shall apply the following procedure to the rows $t_m, m > l$ of the matrix

$$(x - a_i)^C \begin{pmatrix} Y(x) \\ W(x) \end{pmatrix}.$$

If $t_m \equiv 0$, then we stop the procedure (i.e. t_m is not changed at all). Otherwise $t_m = (x - a_i)^{k_m} w_m(x)$, where $w_m(a_i) \neq 0$. If $k_m > \min(c_1, \ldots, c_l)$, then stop the procedure. If $k_m \leq \min(c_1, \ldots, c_l)$, then do the following. Since the rows $y_1(a_i), \ldots, y_l(a_i)$ of the matrix $Y(a_i)$ are linearly independent, we have $w_m(a_i) = -\sum_{j=1}^{l} d_j y_j(a_i)$. Hence, the row vector

$$t_m^1(x) = d_1(x - a_i)^{c_m - c_1} t_1(x) + \ldots + d_l(x - a_i)^{c_m - c_l} t_l(x) + t_m(x) \quad (4.1.4)$$

has the form $t_m^1(x) = (x - a_i)^{c_m^1} w_m^1(x)$, where either $w_m^1(x) \equiv 0$ or $w_m^1(a_i) \neq 0$, $c_m^1 > c_m$. If $w_m^1(x) \equiv 0$ or $w_m^1(a_i) \neq 0$, $c_m^1 > \min(c_1, \ldots, c_l)$, then we stop the procedure. If $c_m^1 \leq \min(c_1, \ldots, c_l)$ and $w_m^1(a_i) \neq 0$, then $w_m^1(a_i) = -\sum_{j=1}^{l} d_j^1 y_j(a_i)$ and we again can consider the corresponding polynomial

$$t_m^2(x) = d_1^1(x - a_i)^{c_m^1 - c_1} t_1(x) + \ldots + d_l^1(x - a_i)^{c_m^1 - c_l} t_l(x) + t_m^1(x)$$

and so on.

In all cases after a finite number of steps, we get $t_m^s(x) = (x - a_i)^{c_m'} w_m'(x)$, where $c_m' > \min(c_1, \ldots, c_l)$ with holomorphic $w_m'(x)$. We consider the polynomials

$$Q_j^m = d_j(x - a_i)^{c_m - c_1} + d_j^1(x - a_i)^{c_m^1 - c_1} + \ldots + d_j^{s-1}(x - a_i)^{c_m^{s-1} - c_1}$$

in $\frac{1}{x - a_i}$. By construction,

$$\sum_{j=1}^{l} Q_j^m t_j(x) + t_m(x) = (x - a_i)^{c_m'} w_m'(x), \quad m = l + 1, \ldots, p.$$

One should substitute

$$\Gamma(x) = \begin{pmatrix} 1 & & & & & & & & \\ & 0 & & \cdot & & & & 0 & \\ & & \cdot & & \cdot & & & & \\ & & \cdot & & & \cdot & & & \\ Q_1^{l+1} & \cdot & \cdot & \cdot & Q_l^{l+1} & 1 & & & \\ & \cdot & & & & & 0 & \cdot & \\ & \cdot & & \cdot & \cdot & & & \cdot & \\ Q_1^p & \cdot & \cdot & \cdot & Q_l^p & 0 & & \cdot & 1 \end{pmatrix} \quad (4.1.5)$$

for the matrix $\Gamma(x)$ in (4.1.3). This concludes the proof of the lemma.

The following corollary of the lemma will be used in Section 5.

Corollary 4.1.1 *Under the assumptions of Lemma 4.1.1 there exists a matrix $\Gamma(x)$ such that, for any row $w'_m(x)$ of the matrix $W'(x)$ in (4.1.3), one of the following additional conditions hold:*

(a) $w'_m(x) \equiv 0$;

(b) $w'_m(a_i) = -\sum_{r=1}^k d_{j_r} y_{j_r}(a_i)$, where $\prod_{r=1}^k d_{j_r} \neq 0$ and $\min(c_{j_1}, \ldots, c_{j_k}) = c_{j_*} < c'_m$.

Proof. If for row $w'_m(x)$ the inequality $c_{j_*} \geq c'_m$ holds, then we apply again procedure (4.1.4), etc. , until we get either (a) or (b); after that we construct the corresponding Q_j^m from (4.1.5). $\blacksquare$

Remark 4.1.1 *It follows from the form (4.1.5), (4.1.4) of $\Gamma(x)$ that for any holomorphic at a_i matrix $Z(x)$ of the size (p, m) the matrix*

$$\Gamma(x)(x - a_i)^C Z(x)$$

is still holomorphic at a_i.

Lemma 4.1.2 *Let a matrix $U(x)$ be holomorphic at a_i and let all the principal minors of $U(a_i)$ be nonzero. Then for any integer-valued diagonal matrix $C = \mathrm{diag}(c_1, \ldots, c_p)$ with the condition $c_1 \geq \ldots \geq c_p$ there exist a holomorphically invertible off a_i matrix $\Gamma(x)$ and a holomorphically invertible in U_i matrix $V(x)$, such that*

$$\Gamma(x)(x - a_i)^C U(x) = V(x)(x - a_i)^C. \qquad (4.1.6)$$

Proof. Rewrite the matrix $(x - a_i)^C U(x)$ as follows:

$$(x - a_i)^C U(x) = (x - a_i)^{C - c_p I} U(x)(x - a_i)^{c_p I} \qquad (4.1.7)$$

and apply Lemma 4.1.1 to the matrix

$$C = C - c_p I, \quad \begin{pmatrix} Y(x) \\ W(x) \end{pmatrix} = \begin{pmatrix} U^{p-1}(x) \\ w_p(x) \end{pmatrix}, \quad l = p - 1,$$

where $U^l(x)$ is formed by the intersections of the rows and columns of $U(x)$ with the numbers $1, \ldots, l$, $w_p(x)$ is the vector function $(u_{p1}, \ldots, u_{pp-1})$, $U(x) = (u_{ij})$.

By Lemma 4.1.1 there exists a matrix $\Gamma_1(x)$ of form (4.1.5) (with $l = p - 1$), such that

$$\Gamma_1(x)(x - a_i)^{C - c_p I} \begin{pmatrix} U^{p-1}(x) \\ w_p(x) \end{pmatrix} = (x - a_i)^{C_1} \begin{pmatrix} U^{p-1}(x) \\ w'_p(x) \end{pmatrix}, \qquad (4.1.8)$$

where $C_1 = \text{diag}(c_1 - c_p, \ldots, c_{p-1} - c_p, c_p^1)$, $c_p^1 > c_{p-1} - c_p$. Therefore

$$(x - a_i)^{C_1} \begin{pmatrix} U^{p-1} \\ w_p' \end{pmatrix} = (x - a_i)^{C_1 - (c_{p-1} - c_p)I} \begin{pmatrix} U^{p-1} \\ w_p' \end{pmatrix} (x - a_i)^{(c_{p-1} - c_p)I}.$$

It follows from (4.1.7) and the latter formula that the following factorization holds:

$$\Gamma_1(x)(x - a_i)^C U(x) = (x - a_i)^{C_1 - (c_{p-1} - c_p)I} \begin{pmatrix} U^{p-2} \\ W_2 \end{pmatrix} | Z_1 \end{pmatrix} (x - a_i)^{D_1}, \quad (4.1.9)$$

where W_2 is a matrix of the size $(2, p - 2)$, W_2, Z_1 are holomorphic at a_i, $D_1 = \text{diag}(c_{p-1}, \ldots, c_{p-1}, c_p)$.

Let apply Lemma 4.1.1 to the matrices

$$C = C_1 - (c_{p-1} - c_p)I, \quad \begin{pmatrix} Y(x) \\ W(x) \end{pmatrix} = \begin{pmatrix} U^{p-2}(x) \\ W_2(x) \end{pmatrix}, \quad l = p - 2.$$

By Lemma 4.1.1 there exists $\Gamma_2(x)$ and so on.

As a result after $p - 1$ steps we obtain a matrix $\Gamma(x) = \Gamma_{p-1} \cdot \ldots \cdot \Gamma_1$, such that (4.1.6) holds with some holomorphic matrix $V(x)$.

Since

$$\det V(a_i) = \lim_{x \to a_i} \det \Gamma(x) \det U(a_i) = \det U(a_i) \neq 0,$$

we obtain that $V(x)$ is holomorphically invertible at a_i. (Here we used form (4.1.5) of each $\Gamma_i(x)$, which implies $\det \Gamma_i(x) \equiv 1$).

Lemma 4.1.3 *Let a matrix $U(x)$ be holomorphically invertible at a_i. Then for any integer-valued diagonal matrix $C = \text{diag}(c_1, \ldots, c_p)$ there exist a holomorphically invertible off a_i matrix $\Gamma(x)$ and a holomorphically invertible at a_i matrix $V(x)$, such that*

$$\Gamma(x)(x - a_i)^C U(x) = V(x)(x - a_i)^D, \quad (4.1.10)$$

where $D = \text{diag}(d_1, \ldots, d_p)$ is obtained by some permutation of diagonal elements of the matrix C.

Moreover, if for some $\Gamma(x)$ having the same properties as before formula (4.1.10) holds with appropriate diagonal matrix D, then the following inequalities :

$$c^1 \leq d^1, \quad c^p \geq d^p. \quad (4.1.11)$$

hold for the elements $c^1 = \max(c_1, \ldots, c_p)$, $d^1 = \max(d_1, \ldots, d_p)$, $c^p = \min(c_1, \ldots, c_p)$, $d_p = \min(d_1, \ldots, d_p)$

Proof. First let the diagonal elements of C be nonincreasing. With help of some constant nondegenerated matrix S we transpose the columns of the matrix $U(x)$ so that all principal minors of the new matrix $U' = US$ are not equal to zero. (S is the matrix of a linear transformation which interchanges vectors of the standard basis of C^p. Note that the conjugation via this matrix transforms a diagonal matrix into a diagonal one). Applying Lemma 4.1.2 to U', we obtain

$$\Gamma(x)(x - a_i)^C U'(x) = V'(x)(x - a_i)^C,$$

therefore ,

$$\Gamma(x)(x - a_i)^C U(x) = \Gamma(x)(x - a_i)^C U'(x)S^{-1} = V'(x)(x - a_i)^C S^{-1} =$$

$$= V'(x)S^{-1}(x - a_i)^{SCS^{-1}} = V(x)(x - a_i)^D.$$

If the elements $c_1, \ldots, c_p$ are not ordered, then there exists a constant matrix S', such that $(S')^{-1}CS' = C'$, where $C' = \mathrm{diag}(c'_1, \ldots, c'_p)$ and $c'_1, \ldots, c'_p$ already form a nonincreasing sequence. For the matrix $(x - a_i)^{C'} S^{-1}U(x)$ consider the corresponding matrix $\Gamma'(x)$. In this case one can take the matrix $\Gamma = \Gamma' S^{-1}$ for the matrices C and $U(x)$ in (4.1.10).

Let us prove the second statement of the lemma. Assume that for some $\Gamma(x)$ from (4.1.10) there exists D with $d^p > c^p$. Let $d^p = d_l,\ c^p = c_m$ for some l, m. It follows from (4.1.10) that

$$\Gamma(x) = V(x)(x - a_i)^D U^{-1}(x)(x - a_i)^{-C}. \qquad (4.1.12)$$

The element u'_{sm} of the matrix $(x - a_i)^D U^{-1} (x - a_i)^{-C}$ has the form $u'_{sm} = u_{sm}(x - a_i)^{d_s - c_m}$, where u_{sm} is holomorphic. Therefore, the m-th column γ_m of the matrix $\Gamma(x)$ has the form $\gamma_m = v_m(x - a_i)^{d_l - c_m}$ with holomorphic column vector $v_m(x)$. Since by the assumption $d^p > c^p$, one has that all elements of γ_m are holomorphic at a_i and vanish there. Then, γ_m is holomorphic on the Riemann sphere off a_i . By Liouville's theorem we obtain $\gamma_m \equiv 0$, which contradicts the holomorphic invertibility of $\Gamma(x)$ outside of a_i.

The proof of the first inequality in (4.1.11) are similar to the proof of the second one. Instead of (4.1.12) one must consider the formula

$$\Gamma^{-1}(x) = (x - a_i)^C U(x)(x - a_i)^{-D} V^{-1},$$

and instead of the column γ_m one must consider the row γ^t of $\Gamma^{-1}(x)$, where $d^1 = d_t$ for some t.

4.2   Solution for an irreducible representation

All counterexamples to Hilbert's 21st problem are constructed for reducible representations. But how does the algebraic property of reducibility appear in analytic theory of ODE? The partial answer on this question is presented in the following lemmas.

Lemma 4.2.1 *If for some component $y_j(\tilde{x})$ of any nonzero function $y(\tilde{x})$ from the space X of solutions of a system (1.2.1) with monodromy χ the identity*

$$y_j(\tilde{x}) \equiv 0$$

holds, then the representation χ is reducible.

Proof. We consider a basis $(e_1(\tilde{x}), \ldots, e_p(\tilde{x}))$ of X, such that

$$e_1(\tilde{x}) = y(\tilde{x}), \quad e_1^j(\tilde{x}) \equiv \ldots \equiv e_l^j(\tilde{x}) \equiv 0,$$

and the functions $e_{l+1}^j(\tilde{x}), \ldots, e_p^j(\tilde{x})$ are linearly independent. It is obvious that l must satisfy the inequality $1 \leq l < p$. Let $m \leq l$ and $\sigma \in \Delta$. Let us consider $\sigma^*(e_m) = \sum_{i=1}^p \lambda_i e_i$. Since, by construction, $e_m^j(\tilde{x}) \equiv 0$ for $1 \leq m \leq l$, it follows that

$$(\sigma^* e_m^j)(\tilde{x}) = e_m^j(\sigma^{-1}\tilde{x}) \equiv 0 = \sum_{i=l+1}^p \lambda_i e_i^j(\tilde{x}) .$$

Since $e_{l+1}^j(\tilde{x}), \ldots, e_p^j(\tilde{x})$ are linearly independent, we have $\lambda_{l+1} = \ldots = \lambda_p = 0$. This means that the subspace $X_l \subset X$ generated by $e_1(\tilde{x}), \ldots, e_l(\tilde{x})$ is the common invariant subspace for the monodromy operators, and so the monodromy representation χ for the system (1.2.1) is reducible.

Corollary 4.2.1 *For any basis $y_1, \ldots, y_p$ of the space of solutions to system (1.2.1) with an irreducible monodromy and for any j the j-components $y_1^j, \ldots, y_p^j$ are linearly independent.*

Proof. Let $c_1 y_1^j + \ldots + c_p y_p^j \equiv 0$ for some $c_1, \ldots, c_p$, $|c_1| + \ldots + |c_p| \neq 0$. Then the solution $c_1 y_1 + \ldots + c_p y_p$ has a zero j-component. Therefore, from Lemma 4.2.1 we have that the monodromy of the system is reducible , that contradicts the assumption of the corollary.

The goal of this subsection consists in proving the following statement.

Theorem 4.2.1 *Any irreducible representation χ can be realized as the monodromy representation of some Fuchsian system.*

Proof. Consider a regular system (1.2.1) with the given monodromy χ, which is Fuchsian outside of the singular point a_1. (The existence of such a system is proved in Section 3.2). For a Levelt's fundamental matrix $Y(\tilde{x})$ of the system we have the following factorization:

$$Y(\tilde{x}) = V_1(x)(x - a_1)^{\Phi} (\tilde{x} - a_1)^{E_1}, \quad \tilde{x} \in \tilde{U}_1^*$$

(see (2.2.25)) with holomorphic $V_1(x)$ and upper-triangular E_1.

Rewrite this factorization as follows:

$$Y(\tilde{x}) = V_1'(x)(x - a_1)^{B} (\tilde{x} - a_1)^{E_1}, \tag{4.2.1}$$

where $B = \mathrm{diag}(b_1, \ldots, b_p)$, $b_i - b_{i-1} > c > 0$, $i = 1, \ldots, p-1$, V_1' is meromorphic at a_1.

Since $V_1'(x)$ is holomorphically invertible in $U_1^* = U_1 \setminus \{a_1\}$, we obtain from Birkhoff-Grothendieck's theorem (see Section 3.2) that there exist a holomorphically invertible off a_1 matrix $\Gamma_1(x)$ and a holomorphically invertible in U_1 matrix $U(x)$, such that

$$\Gamma_1(x)V_1'(x) = (x - a_1)^{C} U(x), \tag{4.2.2}$$

where C is an integer-valued diagonal matrix $C = \mathrm{diag}(c_1, \ldots, c_p)$, $c_1 \geq \ldots \geq c_p$.

Indeed, we may regard $V_1'(x)$ as the transition function of some vector bundle on $\bar{\mathbb{C}}$ with the coordinate neighborhoods $\bar{\mathbb{C}} \setminus \{a_1\}$ and U_1. Formula (4.2.2) means that this bundle is holomorphically equivalent to a bundle with transition function $(x - a_1)^{C}$, and this is exactly the statement of Birkhoff-Grothendieck's theorem (see 3.2.3).

By Lemma 4.1.3 there exists a holomorphically invertible off a_i matrix $\Gamma(x)$, such that

$$\Gamma(x)(x - a_1)^{C} U(x) = V(x)(x - a_1)^{D}, \tag{4.2.3}$$

where D is obtained by a permutation of diagonal elements of C, and $V(x)$ is holomorphically invertible at a_1.

Introduce a new dependent variable $z = \Gamma(x)\Gamma_1(x)y$. From (4.2.1) - (4.2.3) it follows that for the fundamental matrix $Z(\tilde{x}) = \Gamma\Gamma_1 Y(\tilde{x})$ of the new system the following factorization holds:

$$Z(\tilde{x}) = V(x)(x - a_1)^{D+B} (\tilde{x} - a_1)^{E_1}. \tag{4.2.4}$$

If for elements $c_1, \ldots, c_p$ of the matrices C and D the condition

$$c_1 - c_p \le c \tag{4.2.5}$$

were fulfilled, then the diagonal elements of the matrix $D + B$ would form a nonincreasing sequence and by Corollary 2.2.1 we would obtain that the new system were Fuchsian at a_1 too. Indeed, in this case

$$(x - a_1)^{D+B} \, E_1 (x - a_1)^{-D-B}$$

would be holomorphic at a_1 (see the formula below (2.2.27)) and we could apply Corollary 2.2.1.

So all we need for proving of the theorem is the following statement.

Proposition 4.2.1 *Let a regular system (1.2.1) with an irreducible monodromy be Fuchsian outside of the singular point a_1. Then for any matrix B from factorization (4.2.1) of the fundamental matrix $Y(\tilde{x})$ of this system the inequality*

$$\sum_{i=1}^{p} (c_1 - c_i) \le \frac{(n-2)p(p-1)}{2} \tag{4.2.6}$$

is satisfied , where the numbers c_i are from factorization (4.2.2) , and the matrix $V_1'(x)$ is from (4.2.1).

Indeed, if (4.2.6) holds true, then

$$c_1 - c_p \le \sum_{i=1}^{p} (c_1 - c_i) \le \frac{(n-2)p(p-1)}{2} \, ,$$

so one can take $c = \frac{1}{2}(n-2)p(p-1)$ in (4.2.5).

Proof. The differential 1-form $\operatorname{tr}\omega = d \ln \det Y(\tilde{x})$ is a single-valued meromorphic form on $\bar{\mathbb{C}}$ (see (2.2.40)). Since system (1.2.1) is Fuchsian at the points $a_2, \ldots, a_n$ and by virtue of factorization (2.2.45), the residues of the form at the indicated points are

$$\operatorname{res}_{a_i} \operatorname{tr}\omega = \operatorname{tr}\Phi_i + \operatorname{tr}E_i \, , \quad i = 2, \ldots, n \, . \tag{4.2.7}$$

Since $\det \Gamma_1(x) = \text{const} \ne 0$ in (4.2.2) (indeed, by (4.2.2) this function is meromorphic at a_1, because $(V_1')^{-1}$ is meromorphic and U is holomorphic there, and this function is holomorphic and does not vanish anywhere on $\bar{\mathbb{C}} \setminus \{a_1\}$; thus, it can not have neither zero nor pole at a_1, therefore by Liouville's theorem it is nonzero constant), the residue of the form $\operatorname{tr}\omega$ at the pont a_1 is

$$\mathrm{res}_{a_1} \mathrm{tr}\omega = \mathrm{tr}C + \mathrm{tr}B + \mathrm{tr}E_1 \tag{4.2.8}$$

by virtue of (4.2.1), (4.2.2). According to the theorem on the sum of residues, we found from (4.2.7), (4.2.8) that

$$\mathrm{tr}C + \mathrm{tr}B + \sum_{i=2}^{n} \mathrm{tr}\Phi_i + \sum_{i=1}^{n} \mathrm{tr}E_i = 0 . \tag{4.2.9}$$

Let us consider the first row $(y_1, \ldots, y_p)$ of the matrix $\Gamma_1(x)Y(\tilde{x})$, where $\Gamma_1(\tilde{x})$ is that appearing in (4.2.2). It follows from (2.2.45), (4.2.1), (4.2.2) that

$$(y_1, \ldots, y_p) = (u_1^i, \ldots, u_p^i)(x - a_i)^{\Phi_i} (\tilde{x} - a_i)^{E_i} S_i , \quad i = 2, \ldots, n , \tag{4.2.10}$$

$$(y_1, \ldots, y_p) = (v_1, \ldots, v_p)(x - a_1)^{B+c_1 I} (\tilde{x} - a_1)^{E_1} ,$$

where $(u_1^i, \ldots, u_p^i)$ is the first row of the holomorphic matrix $\Gamma_1(x)V_i(x)$, $(v_1, \ldots, v_p)$ is the first row of the holomorphic matrix $V(x)$. (In general, the matrix $\Gamma_1 Y$ is not a Levelt's fundamental matrix in $\tilde{U}_i^*$, but of course, $\Gamma_1 Y$ can be transformed to it by multiplying on a suitable S_i).

Let us consider a matrix $T(\tilde{x})$ whose jth row t^j is

$$t^j = \left(\prod_{i=1}^{n}(x - a_i)^{j-1} \right) \frac{d^{j-1} y}{dx^{j-1}} , \tag{4.2.11}$$

where $y(x)$ denotes the row $y = (y_1, \ldots, y_p)$.

Since by the hypothesis the representation χ of the monodromy of system (1.2.1) is irreducible, the analytic functions $y_1, \ldots, y_p$ are linearly independent (see Corollary 4.2.1, applied to (1.2.1) with the fundamental matrix $\Gamma_1 Y$), therefore the determinant $\det T(\tilde{x})$ is not identically zero on $\bar{\mathbb{C}}$. Its singular points are $a_1, \ldots, a_n$ and the point ∞. Possibly, there are complementary points $\hat{b}_1, \ldots, \hat{b}_m$ at which $\det T(\tilde{x})$ vanishes. Note that the points ∞ and $\hat{b}_1, \ldots, \hat{b}_m$ are points without ramification for $\det T(\tilde{x})$, since the monodromy of the matrix $T(\tilde{x})$ coincides with that of the original system (1.2.1).

Let us consider the form $\gamma = d \ln \det T(\tilde{x})$ and find its residues. According to what was indicated above,

$$d_j = \mathrm{res}_{\hat{b}_j} \gamma \geq 0 , \quad j = 1, \ldots, m . \tag{4.2.12}$$

Let us check that (4.2.10) , (4.2.11) imply that the matrix $T(\tilde{x})$ admits of the factorization

$$T(\tilde{x}) = U_i'(x)(x - a_i)^{\Phi_i} (\tilde{x} - a_i)^{E_i} S_i , \quad i = 2, \ldots, n \tag{4.2.13}$$

in the neighborhood $\tilde{U}_i^*$, where $U_i'(x)$ is a matrix holomorphic at a_i.

Since

$$
t^j = \left(\prod_{l=1, l \neq i}^{n} (x - a_l)^{j-1} \right) (x - a_i)^{j-1} \frac{d^{j-1} y}{dx^{j-1}} ,
$$

it is sufficient to show that

$$
\frac{d^{j-1} y}{dx^{j-1}} = \frac{u^i}{(x - a_i)^{j-1}} (x - a_i)^{\Phi_i} (\tilde{x} - a_i)^{E_i} S_i ,
$$

where u^i is a row vector holomorphic at a_i.

We shall prove this statement by induction. For $j = 1$ it is equivalent to factorization (4.2.10). Suppose that this statement is true for $j = k$. We have

$$
\frac{d^k y}{dx^k} = \frac{d}{dx} \left(\frac{u^i}{(x - a_i)^{k-1}} (x - a_i)^{\Phi_i} (\tilde{x} - a_i)^{E_i} S_i \right) =
$$

$$
= \frac{1}{(x - a_i)^k} \left[(x - a_i) \frac{du^i}{dx} - (k-1)u^i + u^i (\Phi_i + (x - a_i)^{\Phi_i} \cdot \right.
$$

$$
\left. \cdot E_i (x - a_i)^{-\Phi_i}) \right] (x - a_i)^{\Phi_i} (\tilde{x} - a_i)^{E_i} S_i .
$$

Since the expression in square brackets is holomorphic at the point a_i, the statement is true for $j = k$. This completes the proof.

In the similar way we can prove that the factorization

$$
T(\tilde{x}) = V'(x)(x - a_1)^{B + c_1 I} (\tilde{x} - a_1)^{E_1} S_1 \tag{4.2.14}
$$

with the holomorphic matrix $V'(x)$ holds true in the neighborhood of the point a_1. In this case, instead of the holomorphy of the matrix $(x - a_i)^{\Phi_i} E_i (x - a_i)^{-\Phi_i}$, we must use that of the matrix $(x - a_1)^{B + c_1 I} E_1 (x - a_1)^{-B - c_1 I}$ which follows from the fact that $B + c_1 I$ is a diagonal matrix with nonincreasing diagonal elements $b_j + c_1$, $j = 1, \dots, p$ and E_1 is an upper-triangular matrix.

From (4.2.13) we get

$$
\operatorname{res}_{a_i} \gamma = s_i + \operatorname{tr}\Phi_i + \operatorname{tr}E_i, \quad i = 2, \dots, n, \tag{4.2.15}
$$

where $s_i \geq 0, i > 1$ is the order of zero of $\det U_i'(x)$ at a_i and from (4.2.14) we get

$$
\operatorname{res}_{a_i} \gamma = s_1 + \operatorname{tr}B + pc_1 + \operatorname{tr}E_1
$$

with $s_1 \geq 0$, where s_1 is the order of zero of $\det V'(x)$ at a_1.

It remains to find the residue of the form γ at the point ∞. We denote by $W(x)$ the matrix $(\frac{d^{j-1}y_i}{dx^{j-1}})$ and by $R(\zeta)$ the matrix $(\frac{d^{j-1}y_i}{d\zeta^{j-1}})$, $i, j = 1, \ldots, p$, where $\zeta = \frac{1}{x}$. Since the functions $y_1, \ldots, y_p$ are analytic at the point ∞, the same is true for all $R(\zeta)$. Since

$$\frac{d^{j-1}y}{dx^{j-1}} = \frac{1}{x^{2(j-1)}} \sum_{i=2}^{j} k_i^j x^{j-i} \frac{d^{i-1}y}{d\zeta^{i-1}}, \tag{4.2.16}$$

where $k_j^j = (-1)^{j-1}$, it follows that

$$W(x) = \tilde{\Gamma}_2(x) R(\frac{1}{x}), \tag{4.2.17}$$

where, in turn, $\tilde{\Gamma}_2(x) = (\gamma_{ij})$ is a lower-triangular matrix with elements $\gamma_{11} = 1$, $\gamma_{jj} = \frac{k_j^j}{x^{2(j-1)}}$, and therefore

$$\det W(x) = \prod_{j=1}^{p} \frac{k_j^j}{x^{2(j-1)}} \det R(\frac{1}{x}). \tag{4.2.18}$$

Since, according to (4.2.11), we have $T(\tilde{x}) = \tilde{\Gamma}_3(x) W(\tilde{x})$,

$$\tilde{\Gamma}_3(x) = \operatorname{diag}\left(1, \prod_{i=1}^{n}(x - a_i), \ldots, \prod_{i=1}^{n}(x - a_i)^{p-1}\right)$$

from (4.2.18) we find that

$$\det T(\tilde{x}) = x^{\frac{1}{2}(n-2)p(p-1)} \cdot \det R(\zeta) \cdot h(\zeta),$$

where $|h(0)| = 1$. Consequently,

$$\operatorname{res}_\infty \gamma = -\frac{(n-2)p(p-1)}{2} + d', \tag{4.2.19}$$

where d' is the order of the zero of the function $\det R(\zeta)$ for $\zeta = 0$.

According to the theorem on the sum of residues for the form γ and from (4.2.12), (4.2.15), (4.2.19) we find that

$$pc_1 + \operatorname{tr}B + \sum_{i=2}^{n} \operatorname{tr}\Phi_i + \sum_{i=1}^{n} \operatorname{tr}E_i + d = \frac{(n-2)p(p-1)}{2}, \tag{4.2.20}$$

where $d = \sum_{i=1}^{m} d_i + \sum_{i=1}^{n} s_i + d_i' \geq 0$.

Subtracting (4.2.9) from (4.2.20) and discarding $d \geq 0$, we get the required inequality (4.2.6).

Remark 4.2.1 *The only place in the proof where we used the irreducibility of the representation χ is the statement concerning the linear independence of the elements $y_1, \ldots, y_p$ of the first row of the matrix $\Gamma_1(x)Y(\tilde{x})$.*

An independent proof of Theorem 4.2.1 was obtained by V.Kostov [Ko1], [Ko2].

The inequality (4.2.6) has some geometric sense, which will be explained below.

5 Miscellaneous topics

5.1 Vector bundles and Hilbert's 21st problem

An answer to Hilbert's 21st problem can be expressed in terms of holomorphic sections of some vector bundles over $\bar{\mathbb{C}}$.

Let us again consider the bundles E, P, constructed in Section 3.2. Let $G_i = \chi(\sigma_i^{-1})$, $E_i = \frac{1}{2\pi i} \ln G_i$ (see Section 3.2). With help of a constant matrix S_i transform E_i to an upper-triangular form $E_i^0 = S_i^{-1} E_i S_i$. Consider a matrix

$$\Lambda_i = \text{diag}(\lambda_1, \ldots, \lambda_p), \quad \lambda_i \in \mathbb{Z}, \quad \lambda_1 \geq \cdots \geq \lambda_p.$$

Denote by λ^i the collection $(\lambda_1, \ldots, \lambda_p)$ and call it *admissible*.

Let us replace the glueing functions ω_i^E, ω_i^P from (3.2.1) by $\omega_{\lambda^i}^E$, $\omega_{\lambda^i}^P$ as follows. Replace f_i from (3.2.2) by the following one:

$$f_{\lambda^i}(\tilde{x}, \sigma) = S_i (x - a_i)^{\Lambda_i} (\tilde{x} - a_i)^{E_i^0} S_i^{-1} \chi(\sigma^{-1}) \tag{5.1.1}$$

and replace $g_i(\tilde{x})$ by $g_{\lambda^i}(\tilde{x}) = f_{\lambda^i}(\tilde{x}_0, \sigma)$ (see the notation of (3.2.1)).

Let us do such a procedure for all $a_1, \ldots, a_n$. As a result we obtain bundles G^λ and R^λ, $\lambda = (\lambda^1, \ldots, \lambda^n)$ on $\bar{\mathbb{C}}$.

Remark 5.1.1 *In general such extensions of the original vector bundles depend not only on a collection of λ but also on a choice of matrices S_i. But we shall not mark this fact in notation of extensions (having it in mind).*

Theorem 5.1.1 *A representation χ can be realized as the monodromy representation of some Fuchsian system if and only if there exists an admissible collection $\lambda = (\lambda^1, \ldots, \lambda^n)$ such that the corresponding principal bundle R^λ (and the associated vector bundle G^λ) is holomorphically trivial.(Cf. [Bo2], [Bo5]).*

Proof. The proof immediately follows from the Levelt theorem (Theorem 2.2.1). Indeed, let R^λ with some admissible $\lambda = (\lambda^1, \ldots, \lambda^n)$ (which means that each λ^i is admissible) be holomorphically trivial. Denote by $W(x)$ a holomorphic section of the bundle. Consider again the "multivalued" section $Z(\tilde{x})$ from (3.1.5) and the corresponding matrix function $Y : \tilde{S} \longrightarrow \text{GL}(p; \mathbb{C})$, where

$$Z(\tilde{x}) = W(p(\tilde{x}))Y(\tilde{x}). \tag{5.1.2}$$

It was proved in Section 3.1 that $Y(\tilde{x})$ was the fundamental matrix for system (1.2.1) with the given monodromy and singular points. In local coordinates $\Phi_i^{R^\lambda}$ ($\Phi_i^{R^\lambda}$ corresponds to $\omega_{\lambda_i}^P$, cf. Section 3.2), we have

$$Y(\tilde{x}) = (p_2 \Phi_i^{R^\lambda}(i_P W(x)))^{-1}(p_2 \Phi_i^{R^\lambda}(i_P Z(\tilde{x}))), \; (x = p\tilde{x}) \tag{5.1.3}$$

($i_P : P \to R^\lambda$ is analogous to $i_P : P \to R$ from Section 3.2; we do not mark in the notation that now i_P depends on λ and $\{S_i\}$), where the first factor is holomorphic at a_i and by the construction the second factor has the next form for $\tilde{x} = \sigma\tilde{x}_0, \tilde{x}_0 \in \tilde{U}_i^*$:

$$\Phi_i^{R^\lambda}(i_P Z(\tilde{x})) = \omega_{\lambda_i}^P(Z(\tilde{x})) = \omega_{\lambda_i}^P(<\tilde{x}, I>) =$$

$$= \omega_{\lambda_i}^P(<\sigma\tilde{x}_0, I>) = \omega_{\lambda_i}^P(<\tilde{x}_0, \chi(\sigma^{-1})>) = (x, g_{\lambda_i}(\tilde{x}_0)\chi(\sigma^{-1})) =$$

$$= (x, S_i(x - a_i)^{\Lambda_i}(\tilde{x}_0 - a_i)^{E_i^0} S_i^{-1}\chi(\sigma^{-1})), \tag{5.1.4}$$

$$p_2\Phi_i^{R^\lambda}(i_P Z(\tilde{x})) = S_i(x - a_i)^{\Lambda_i}(\tilde{x}_0 - a_i)^{E_i^0} S_i^{-1}\chi(\sigma^{-1}).$$

(Cf. the similar calculation in Section 3.2).

Therefore the fundamental matrix $Y'(\tilde{x}) = Y(\tilde{x})S_i$ of the system has the following form in $\tilde{U}_i^*$:

$$Y'(\tilde{x}) = V_i(x)(x - a_i)^{\Lambda_i}(\tilde{x} - a_i)^{E_i^0}, \tag{5.1.5}$$

where $V_i(x) = (p_2\Phi_i^{R^\lambda}(i_p W(x)))^{-1}S_i$ is a matrix holomorphically invertible at a_i, Λ_i is the integer-valued diagonal matrix with nonincreasing diagonal elements and E_i^0 is upper-triangular. Again by formula below (2.2.27) it follows that the matrix $(x - a_i)^{\Lambda_i}E_i^0(x - a_i)^{-\Lambda_i}$ is holomorphic at a_i. Therefore by Corollary 2.2.1 we get that system (1.2.1) is Fuchsian at a_i. So if some bundle R^λ is holomorphically trivial, then the corresponding representation χ can be realized as the monodromy representation of some Fuchsian system.

Now let us assume that for the representation χ the problem has a positive solution. Consider a fundamental matrix $Y(\tilde{x})$ of Fuchsian system (1.2.1) with the given χ. Let $Y_i(\tilde{x})$ be a Levelt's fundamental matrix for the system. Then $Y_i(\tilde{x}) = Y(\tilde{x})S_i$ for some S_i. Consider the admissible collection $\lambda^i = (\lambda_1, \ldots, \lambda_p)$ with $\lambda_j = \varphi_i^j$, where $\Phi_i = \mathrm{diag}(\varphi_i^1, \ldots, \varphi_i^p)$ is from factorization (2.2.25) for the matrix $Y_i(\tilde{x})$. Extend the bundle P at a_i by (5.1.1) with $\Lambda_i = \Phi_i$ and with $S_i = Y^{-1}(\tilde{x})Y_i(\tilde{x})$. In the similar way extend this bundle at other singular points. As a result we get the bundle R^λ with $\lambda = (\lambda^1, \ldots, \lambda^n)$.

The section $W(x)$ of the bundle R^λ, defined by (5.1.2) is holomorphic at each a_i. Indeed, by (5.1.1) and (5.1.2.) we have

$$p_2 \Phi_i^{R^\lambda}(i_P W(x)) = (p_2 \Phi_i^{R^\lambda}(i_P Z(\tilde{x}))) Y^{-1}(\tilde{x}) =$$

$$= S_i (x - a_i)^{\Lambda_i}(\tilde{x}_0 - a_i)^{E_i^0} S_i^{-1} \chi(\sigma^{-1})(V_i(x)(x - a_i)^{\Phi_i}(\tilde{x}_0 - a_i)^{E_i^0} \cdot$$

$$\cdot S_i^{-1} \chi(\sigma^{-1}))^{-1} = S_i V_i(x)^{-1}, \ \tilde{x} = \sigma \tilde{x}_0, \ \tilde{x}_0 \in \tilde{U}_i^*$$

and this matrix is holomorphically invertible at a_i. (A point $\tilde{x}$ has not to lie in $\tilde{U}_i^*$, therefore the factor $\chi(\sigma^{-1})$ appears here, cf. Section 2.2. Computing Y^{-1}, we take into account that the monodromy for Y_i is $S_i^{-1} \chi S_i$).

So $W(x)$ is holomorphic throughout all $\bar{\mathbb{C}}$ and we get that the bundle R^λ is holomorphically trivial.

As we marked above (cf. Section 1.2) instead of a holomorphic section of P (and instead of Grauert's theorem on Stein manifolds) one can deal with a holomorphic connection ∇ in E and P. As it is well known, for E and P such a connection always exists (cf. e.g. [At]) and has the given monodromy.

The extension R^0 of P with $\lambda = (0, \dots, 0)$, described in Section 3.2 is called *the canonical extension* (this extension does not depend on the choice of S_i). At first this extension was considered by Nastold in [Na]. In the case of n variables it was described in [Del]. This extension provides at most logarithmic singularities for the connection ∇ at the points a_i. Above, using Levelt's theorem, we described all extensions possessing such a property.

As we were informed by P.Deligne, all such extensions (in the case of several complex variables) in terms of a so-called splittable filtration were described in [EsVi].

Let us consider the vector bundle G^λ, associated with R^λ. By Birkhoff-Grothendieck's decomposition (3.2.3) we have

$$G^\lambda \cong \mathcal{O}(c_1^\lambda) \oplus \cdots \oplus \mathcal{O}(c_p^\lambda), \ c_1^\lambda \geq, \cdots \geq c_p^\lambda. \tag{5.1.6}$$

We shall say that the number

$$\gamma(\lambda) = \sum_{j=1}^{p}(c_1^\lambda - c_j^\lambda).$$

is *the weight of the bundle G^λ.*

The following statement is an easy corollary of the previous theorem.

Proposition 5.1.1 *A representation χ can be realized as the monodromy representation of some Fuchsian system if and only if there exists an admissible collection $\lambda = (\lambda^1, \ldots, \lambda^n)$ such that the weight $\gamma(\lambda)$ of the corresponding vector bundle G^λ is equal to zero.*

Proof. If there exists a Fuchsian system with the given monodromy χ and singular points $a_1, \ldots, a_n$, then by Theorem 5.1.1 the corresponding vector bundle G^λ is holomorphically trivial. Therefore $c_1^\lambda = \cdots = c_p^\lambda = 0$ in its decomposition (5.1.6). Thus, $\gamma(\lambda) = 0$ for this bundle.

If for some admissible collection $\lambda = (\lambda^1, \ldots, \lambda^n)$ the number $\gamma(\lambda)$ equals zero for a vector bundle G^λ, then $c_1^\lambda = \cdots = c_p^\lambda = c$ in (5.1.4) for its decomposition into line bundles. Consider the bundle $R^{\tilde\lambda}$, where

$$\tilde\lambda = (\tilde\lambda^1, \ldots, \lambda^n), \quad \tilde\lambda^1 = (\lambda_1 + c, \ldots, \lambda_p + c).$$

Let us prove that this bundle is holomorphically trivial.

Let $W(x)$ be a meromorphic section of R^λ, holomorphic off a_1, such that

$$p_2 \Phi_1^{R^\lambda}(i_P W(x)) = V(x)(x - a_1)^{-cI}$$

with a holomorphically invertible $V(x)$ in a neighborhood U_1 of the point a_1. Such a section always exists. Indeed, under conditions of the proposition there exist $\tilde V(x)$ and $V(x)$, such that $\tilde V(x)$ is holomorphically invertible in $\bar{\mathbb C} \setminus \{a_1\}$, $V(x)$ is holomorphically invertible in U_1 and

$$\tilde V(x)(x - a_1)^{cI} V^{-1}(x) = g_{\infty 1}(x).$$

(This statement is equivalent to the Birkhoff-Grothendieck theorem). Now one can take the section $W(x)$, whose coordinate description is as follows: $\tilde V(x)$ in $\bar{\mathbb C} \setminus \{a_1\}$ and $V(x)(x - a_1)^{-cI}$ in coordinates $\Phi_1^{R^\lambda}$ in U_1.

Consider again the bundle $R^{\tilde\lambda}$. From (5.1.3) and the fact that a scalar matrix commutes with other matrices we obtain

$$p_2 \Phi_1^{R^{\tilde\lambda}}(i_P W) = (x - a_1)^{cI} p_2 \Phi_1^{R^\lambda}(i_P W)$$

(here left i_P is $P \to P^{\tilde\lambda}$, right i_P is $P \to P^\lambda$). Therefore $p_2 \Phi_1^{R^{\tilde\lambda}}(i_P W(x)) = V(x)$ and we get that the same section $W(x)$, regarded as the section of the bundle $R^{\tilde\lambda}$ is already holomorphic over whole $\bar{\mathbb C}$. Thus, $R^{\tilde\lambda}$ (and $G^{\tilde\lambda}$) is holomorphically trivial . Hence, the statement follows from Theorem 5.1.1.

Consider all extensions G^λ for all admissible λ and consider weights $\gamma(\lambda)$.

We say that the number $\gamma_m(\chi) = \sup_\lambda \gamma(\lambda)$ is *the maximal Fuchsian weight of the representation* χ.

Splitting type $(c_1, \ldots, c_p)$ of the bundle G^λ and its weight can be expressed in terms of Levelt's factorization of the corresponding regular system (1.2.1). Consider a section $W(x)$ of R^λ such that the following conditions hold:

i) $W(x)$ is holomorphic off a_1;

ii) $p_2 \Phi_1^{R^\lambda}(i_P W(x)) = V(x)(x - a_1)^{-C}$,
where $C = \mathrm{diag}(c_1, \ldots, c_p)$. (The existence of such a section was proved in the second part of Proposition 5.1.1). Then the fundamental matrix $Y(\tilde{x})$ of the corresponding system (1.2.1) has the form

$$Y(\tilde{x}) = (x - a_1)^C V^{-1}(x) S_1 (x - a_1)^{\Lambda_1} (\tilde{x}_0 - a_1)^{E_0} S_1^{-1} \chi(\sigma^{-1}) \qquad (5.1.7)$$

for $\tilde{x} \in p^{-1}(U_i^*)$, $\tilde{x} = \sigma \tilde{x}_0$, $\sigma \in \Delta$.

This system is Fuchsian at $a_2, \ldots, a_n$ with the matrices of valuations $\Phi_2 = \Lambda_2, \ldots, \Phi_n = \Lambda_n$. And its nonfuchsian part $(x - a_1)^C$ defines the splitting type of G^λ.

From this point of view the first part of the proof of Theorem 4.2.1 (until Proposition 4.2.1) is closed to the following statement.

Theorem 5.1.2 *If the maximal Fuchsian weight of the representation χ is bounded, then χ can be realized as the monodromy representation of some Fuchsian system.*

Proof. We may assume that $a_1 = 0$ and E_1 is upper-triangular. Consider R^λ with $\lambda = (\lambda^1, 0, \ldots, 0)$, $\lambda^1 = (b_1, \ldots, b_p)$, $S_1 = I$ and apply Birkhoff-Grothendieck's theorem to the transition function $g_{\infty 1}$ (we use coordinates $\Phi_1^{R^\lambda}$ over neighborhood U_1 of $a_1 = 0$ and some coordinates which we do not specify in $U_\infty = \mathbb{C}$). We get $g_{\infty 1} = V_\infty x^C V_1$. Denote by W the section $\bar{\mathbb{C}} \setminus \{0\} \to R^\lambda$ having coordinates V_∞ over U_∞; its coordinates over U_1 are $V_1^{-1} x^{-C}$. Write our standard "branching cross-section" Z as $Z = WY$. An easy computation reveals that $Y = x^C V_1 x^B \tilde{x}^{E_1}$ in $\tilde{U}_1^*$. Applying Lemma 4.1.3 to V_1 we get a Γ such that $\Gamma x^C V_1 = V x^D$, Γ is holomorphically invertible outside of a_1, V is holomorphically invertible in U_1 and D is a diagonal matrix whose coefficients are c_j up to order. Thus $\Gamma Y = V x^{B+D} \tilde{x}^{E_1}$ is the solution to a system which is Fuchsian at a_1 if the diagonal coefficients of $B + D$ are in a nonincreasing order. Finiteness of $\gamma_m(\lambda)$ guarantees that this will be the case for appropriate B. Also $Z = (W\Gamma^{-1})(\Gamma Y)$ and $W\Gamma^{-1}$ is the cross-section which is holomorphically invertible outside of a_1, so the corresponding system will be Fuchsian at $a_2, \ldots, a_n$. Clearly ΓY has the same monodromy as Y, i.e. as Z.

Mark here that inequality (4.2.5) in the proof of Theorem 4.2.1 is exactly the corollary of the condition of finiteness for the maximal Fuchsian weight of χ. And Proposition 4.2.1 is equivalent to the following one.

Proposition 5.1.2 *The maximal Fuchsian weight $\gamma_m(\chi)$ of an irreducible representation is bounded as follows:*

$$\gamma_m(\chi) \leq \frac{(n-2)p(p-1)}{2}.$$

So, the positive answer to Hilbert's 21st problem for an irreducible representation is just the corollary of two latter geometric statements.

5.2   Reducibility and regular systems

In what follows we shall need some properties of system (1.2.1) connected with the reducibility of its monodromy representation χ.

Lemma 5.2.1 *If the matrix A of system (1.2.1) satisfies the condition*

$$a_{ij} \equiv 0, \quad i = l+1,\ldots,p; \quad j = 1,\ldots,l; \quad l < p, \tag{5.2.1}$$

then the monodromy representation χ for the system (1.2.1) is reducible.

Proof. Let us consider the system

$$\frac{df}{dx} = A' f,$$

where $A' = (a_{ij})$ for $1 \leq i,j \leq l$. If f is a solution of the latter system, then the column vector $y = (f, 0, \ldots, 0)$ is a solution of the original system (1.2.1). Lemma 5.2.1 follows now from Lemma 4.2.1.

Lemma 5.2.2 *Let the monodromy representation χ for the system (1.2.1) with regular singular points $a_1,\ldots,a_n$ be reducible and let X_l be a common invariant subspace for the monodromy operators in the space of solutions of the system. Then the sum s_l of the exponents of X_l over all the points $a_1,\ldots,a_n$ is an integer and satisfies the inequality*

$$s_l = \sum_{i=1}^{n}\sum_{j=1}^{l} \beta_i^j \leq 0. \tag{5.2.2}$$

Proof. Let us choose a basis $y_1, \ldots, y_l$ in X_l and let us consider the fundamental matrix $Y(\tilde{x})$ constructed by extending this basis. We denote by $Y'(\tilde{x})$ a matrix consisting of the elements of a base minor of the $(p \times l)$- matrix $(y_1(\tilde{x}), \ldots, y_l(\tilde{x}))$ at $\tilde{x}_0 \in \tilde{S}$ (i.e. nondegenerate l-th order minor of the latter matrix). Then $\det Y'(\tilde{x}_0) \neq 0$. Note here that the monodromy for Y' can be described via multiplication on the right by the left upper $(l \times l)$-block χ_1 of χ. The space X' generated by the columns of $Y'(\tilde{x})$ is the space of solutions to a system of form (1.2.1) with the coefficient matrix $A' = \frac{dY'}{dx}(Y')^{-1}$. The set of singular points of the system consists of $a_1, \ldots, a_n$ and of additional apparent singularities $a'_{n+1}, \ldots, a'_r$. The latter set of singularities contains points such that $\det Y'(\tilde{x}) = 0$ for $\tilde{x} \in p^{-1}(a'_t)$. We remark that if $\det Y'(\tilde{x}) = 0$ for some $\tilde{x} \in p^{-1}(a'_t)$, then by (1.2.7) $\det Y'(\tilde{x}) = 0$ for all $\tilde{x} \in p^{-1}(a'_t)$. These singularities are called *apparent*, because they are not ramification points for solutions of the system.

It follows from this remark that the number of additional singular points is finite, because otherwise the set $\{p^{-1}(a'_{n+i})\}$ has a point of accumulation $\tilde{x}_1 \in \tilde{S}$, or the set $\{a'_{n+i}\}$ has one of the points a_j $(j = 1, \ldots, n)$ as a point of accumulation. In the first case the uniqueness theorem for analytic functions applied to $\det Y'(\tilde{x})$ yields $\det Y'(\tilde{x}) \equiv 0$, which contradicts the condition $\det Y'(\tilde{x}_0) \neq 0$. In the second case $\det Y'(\tilde{x})$ has a monodromy described by the multiplier $\det \chi_1(\sigma)$ and grows at most polynomially when x tends to a_j. Thus

$$x \mapsto \det Y'(\tilde{x})(\det \chi'(\sigma))^{-1}$$

is a single valued analytic function, which has at most a pole at a_j and is not identically zero. Its zeroes also cannot accumulate to a_j.

The exponents $\tilde{\beta}^j_{n+i}$ of the space X' at the points a'_{n+i} coincide with the valuations $\tilde{\varphi}^j_{n+i}$, which in turn are non-negative, since $Y'(\tilde{x})$ is analytic at the points $\{p^{-1}(a'_{n+i})\}$:

$$\tilde{\beta}^j_{n+i} = \tilde{\varphi}^j_{n+i} \geq 0. \tag{5.2.3}$$

The valuations $\tilde{\varphi}^j_i$ for X' are connected with the valuations φ^j_i for X_l at the points $a_1, \ldots, a_n$ by the inequalities

$$\tilde{\varphi}^j_i = \varphi_i(y'_j) \geq \varphi_i(y_j) = \varphi^j_i , \tag{5.2.4}$$

which follow from Definition 2.2.4 of the valuations and the fact that the column vector $y'_j(\tilde{x})$ of the matrix $Y'(\tilde{x})$ can be obtained from the vector-valued function $y_j(\tilde{x})$ by crossing out some of its components.

From Theorem 2.2.2 and inequalities (5.2.3) and (5.2.4) we obtain

$$s_l = \sum_{i=1}^{n} \sum_{j=1}^{l} \beta_i^j \leq \sum_{i=1}^{n} \sum_{j=1}^{l} \tilde{\beta}_i^j + \sum_{i=n+1}^{r} \sum_{j=1}^{l} \tilde{\varphi}_i^j = \tilde{s}_l \leq 0,$$

where $\tilde{s}_l$ is the sum of exponents of X'.

Theorem 5.2.1 *Suppose that a representation χ is reducible. Let χ_1 be a subrepresentation of dimension l of the representation χ. There exists a system (1.2.1) with the given monodromy χ and regular singular points $a_1, \ldots, a_n$ which is Fuchsian at all points, with the exception of possibly one, such that its fundamental matrix $Y(\tilde{x})$ has the form*

$$Y(\tilde{x}) = \begin{pmatrix} T^1 & * \\ 0 & T^2 \end{pmatrix}, \tag{5.2.5}$$

where the matrix function $T^1(\tilde{x})$ has the size (l, l) and is invariant under the action of the monodromy χ_1 (cf. [Bo5]).

Proof. We consider system (1.2.1) with the given monodromy χ, which is Fuchsian at the points $a_2, \ldots, a_n$ and with a regular singular point a_1. (The existence of such a system was proved in Section 3.2).

Let $Y(\tilde{x})$ be a fundamental matrix of the space X of solutions of this system such that its first l columns form a Levelt's basis of the l-dimensional subspace $X_l \subset X$ which is invariant under the action of the monodromy χ_1. Then

$$Y(\tilde{x}) = (T_l(\tilde{x})|*). \tag{5.2.6}$$

Due to the beginning of the proof of Theorem 4.2.1 (up the end of Formula (4.2.2)) we may assume that $T_l(\tilde{x})$ has the form:

$$T_l(\tilde{x}) = (x - a_1)^C V^l(x)(x - a_1)^{\Lambda_1^l}(\tilde{x} - a_1)^{E_1^l}, \ \tilde{x} \in \tilde{U}_1^*, \tag{5.2.7}$$

where Λ_1^l is the matrix of valuations of the subspace X_l; $E_1^l = \frac{1}{2\pi i} \ln \chi_1(\sigma_1)$, rank$V^l(a_1) = l$.

Let Y_l be the matrix formed by l rows of the matrix $T_l(\tilde{x})$, with determinant not identically equal to zero. The space X', generated by the columns of the matrix Y_l is the space of solutions of a system of form (1.2.1) with the coefficient matrix $A_l = \frac{dY_l}{dx}Y_l^{-1}$. This system has monodromy χ_1 and singular points $a_1, \ldots, a_n$, as well as, possibly, apparent singular points $a_1', \ldots, a_k'$. These are the points where $\det Y_l(\tilde{x}) = 0$ (see the previous lemma). Hence the sum s' of exponents of the space X' at these points is nonnegative

$$s' \geq 0. \tag{5.2.8}$$

For the sum s'' of exponents of the space X' at the points $a_1, \ldots, a_n$, we have

$$s'' \geq s_l, \qquad (5.2.9)$$

where s_l is the sum of exponents of the space X_l (cf. (5.2.4)).

By the theorem on the sum of residues applied to the form $\mathrm{tr} A_l dx$, we get in the same manner as in Theorem 2.2.2

$$s_c + s'' + s' = 0, \qquad (5.2.10)$$

where s_c is the sum of diagonal elements of the matrix C, which occur in the rows with the same numbers as the rows of Y_l. From this, taking into consideration (5.2.8) and (5.2.9), we have

$$s_c \leq -s_l. \qquad (5.2.11)$$

With no loss of generality, one can assume that Y_l is formed by the first rows of the matrix $T_l(\tilde{x})$. (If this is not the case, we rearrange the rows of the matrix T_l, which corresponds to passing from the matrix T_l to ST_l, where $\det S \neq 0$.) Denote by V_l^l the matrix formed by the rows of V^l with the numbers $1, \ldots, l$. The matrix

$$(x - a_1)^C V^l(\tilde{x})$$

from (5.2.7) has the form required in Lemma 4.1.1 (with $Y(x)$ replaced by V_l^l). By Corollary 4.1.1 there is a matrix $\Gamma(x)$ such that either for all rows of the matrix $W'(x)$ from (4.1.3) condition (a) holds , or for some row w'_m condition (b) is valid. In the latter case, interchanging the rows with numbers j_s and $l + m$ of the matrix

$$(x - a_1)^{C'} \begin{pmatrix} V_l^l(x) \\ W'(x) \end{pmatrix},$$

we obtain the matrix

$$(x - a_1)^{C_1} \begin{pmatrix} \tilde{V}_l^l(x) \\ W''(x) \end{pmatrix}$$

with $\det \tilde{V}_l^l(a_1) \neq 0$.

Under such a transformation the matrix $T_l(\tilde{x})$ transforms to the following one:

$$T_l^1(\tilde{x}) = \Gamma(x)T_l(\tilde{x}) = (x - a_1)^{C_1} \tilde{V}^l(x)(x - a_1)^{\Lambda_1^l}(\tilde{x} - a_1)^{E_1^l}, \ \tilde{x} \in \tilde{U}_1^*. \quad (5.2.12)$$

By condition (b) of Corollary 4.1.1 we have $s_{c_1} > s_c$, where s_{c_1} is the sum of the first l diagonal elements of C_1. Moreover, since $\det \Gamma(x) \neq 0$ for $x \neq a_1$, we get $s'_l \geq s_l$ for the space generated by the columns of T'_l. From (5.2.10) and the latter inequalities we obtain

$$s_c < s_{c_1} \leq -s'_l \leq -s_l. \tag{5.2.13}$$

If for some row of the matrix $W''(x)$ condition (b) holds, then we apply the procedure of Lemma 4.1.1 and Corollary 4.1.1, etc. It follows from (5.2.13) that, after a finite number of steps, we will get the matrix

$$\Gamma_0(x)(x - a_1)^C V^l(x),$$

which satisfies condition (a) of Corollary 4.1.1.

Passing from $Y(\tilde{x})$ of (5.2.6) to $\tilde{Y}(\tilde{x}) = \Gamma_0(x)Y(\tilde{x})$, we get the matrix of form (5.2.5). Since $\det \Gamma_0(x) \neq 0$ for $x \neq a_1$, it follows that $\tilde{Y}(\tilde{x})$ is still the fundamental matrix of a system which is Fuchsian at the points $a_2, \dots, a_n$. The point a_1 is its regular singular point, and there are no other singular points for the system.

Corollary 5.2.1 *Let all monodromy matrices* $G_i = \chi(\sigma_i)$, $i = 1, \dots, n$ *can be simultaneously reduced to the form*

$$G_i = \begin{pmatrix} \underline{G_i^1|} \\ 0 & |\underline{G_i^2} & * \\ & & \ddots \\ & & & \ddots \\ 0 & & & |\underline{G_i^k} \end{pmatrix}, \; i = 1, \dots, n, \tag{5.2.14}$$

where the collection $G_1^j, \dots, G_n^j$ *is irreducible for all* $j = 1, \dots, k$. *Then there exists a regular system (1.2.1) (with the given* χ*) with a fundamental matrix* $Y(\tilde{x})$ *such that the system is Fuchsian off the point* a_1 *and* $Y(\tilde{x})$ *has the form (5.2.14) (with replacing* G_i^j *by* $Y^j(\tilde{x})$*).*

Proof. Denote by χ_l the subrepresentation of χ, defined by the collection

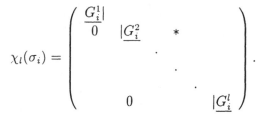

$$\chi_l(\sigma_i) = \begin{pmatrix} \underline{G_i^1|} \\ 0 & |\underline{G_i^2} & * \\ & & \ddots \\ & & & \ddots \\ 0 & & & |\underline{G_i^l} \end{pmatrix}.$$

By Theorem 5.2.1 there exists a system with the given monodromy, whose fundamental matrix $Y(\tilde{x})$ has the form

$$Y(\tilde{x}) = \begin{pmatrix} Y^1(\tilde{x}) & * \\ 0 & \tilde{Y}(\tilde{x}) \end{pmatrix},$$

where $Y^1(\tilde{x})$ is a fundamental matrix of a system with the monodromy χ_1. Then the matrix $\tilde{Y}(\tilde{x})$ is the fundamental matrix of a system (1.2.1) with the representation χ/χ_1. Apply to $\tilde{Y}(\tilde{x})$ Theorem 5.2.1 and so on. By k steps we shall get the matrix $Y'(\tilde{x})$ that we need.

Remark 5.2.1 *The analogous statement for a Fuchsian system is false (see Examples 5.5.1 and 5.5.2).*

The following statement is the corollary of Lemma 4.1.1.

Theorem 5.2.2 *Let a Fuchsian system (1.2.1) have a reducible monodromy χ and let X_l be an invariant under all monodromy operators subspace of the space X of solutions to system (1.2.1). Let the sum s_l of exponents of the space X_l over all singular points $a_1, \ldots, a_n$ be equal to zero. Then there exists a constant nonsingular matrix S such that the matrix A' of the Fuchsian system*

$$\frac{dz}{dx} = A'(x)z, \qquad (5.2.15)$$

obtained from (1.2.1) by changing $z = Sy$ of depending variable y has the form (5.2.1).

Proof. Consider a fundamental matrix $Y(\tilde{x})$ of (1.2.1) such that its first l columns form the basis in X_l. Then by (5.2.6)

$$Y(\tilde{x}) = (T_l(\tilde{x})|*).$$

For the corresponding matrix Y_l from the proof of Theorem 5.2.1 and for the space X', from (5.2.8)—(5.2.10) and from the condition $s_l = 0$ we obtain

$$s_l = s'' = s' = 0.$$

(Since (1.2.1) is Fuchsian, we have $C = 0$ in (5.2.7)). The matrix $V^l(x)$ has the form

$$V^l(x) = \begin{pmatrix} V_l^l(x) \\ W(x) \end{pmatrix}$$

in our case and $\det V_l^l(a_1) \neq 0$, since the corresponding system with the fundamental matrix

$$Y_l^* = V_l^l(x)(x - a_1)^{\Lambda_1^l}(\tilde{x} - a_1)^{E_1^l}, \ \tilde{x} \in \tilde{U}_1^*$$

is Fuchsian (it follows from the condition $s' = 0$ and Theorem 2.2.2.). From latter inequality in Lemma 4.1.1 it follows that there exists a constant matrix $\Gamma(x) = S$ (see (4.1.5)), such that

$$S \left(\begin{array}{c} V_l^l(a_1) \\ W(a_1) \end{array} \right) = \left(\begin{array}{c} V_l^l(a_1) \\ 0 \end{array} \right),$$ (5.2.16)

but this means that

$$SV^l(x) = (x - a_1)^C \left(\begin{array}{c} V_l^l(x) \\ W'(x) \end{array} \right),$$ (5.2.17)

where $C = \mathrm{diag}(0, \ldots, 0, c_{l+1}, \ldots, c_p)$ with $c_j > 0$, $j = l+1, \ldots, p$.

Let for some m the row $w_m'(x)$ of the matrix $W'(x)$ satisfy condition (b) of Corollary 4.1.1, then interchanging the rows with the numbers j_s and $l + m$ of the matrix $SV^l(x)$, we obtain the matrix

$$(x - a_1)^{C_1} \left(\begin{array}{c} \tilde{V}_l^l(x) \\ W''(x) \end{array} \right)$$

with $\det \tilde{V}_l^l(a_1) \neq 0$ (see the corresponding part of the proof of Theorem 5.2.1). As in Theorem 5.2.1 (below formula (5.2.11)) we get $s_{c_1} > s_c = 0$, but $s_{c_1} \leq -s_l = 0$, which contradicts the previous inequality. This means that for all m the matrix $W'(x)$ satisfies condition (a) of Corollary 4.1.1. Hence $W'(x) \equiv 0$ in (5.2.17), therefore the fundamental matrix $SY(\tilde{x}) = Z(x)$ of system (5.2.15) has the form

$$Z(\tilde{x}) = \begin{array}{c} l \\ \end{array} \left(\begin{array}{cc} * & * \\ \overline{0|} & * \end{array} \right)$$
$$\phantom{Z(\tilde{x}) =} \begin{array}{c} \\ l \end{array}$$

and thus, $A'(x)$ has the required form (5.2.1).

Proposition 5.2.1 *Let a Fuchsian system (1.2.1) have a reducible monodromy χ and let each monodromy matrix G_i can be reduced to the Jordan normal form, consisting of only one block. Then for valuations φ_i^j and exponents β_i^j of the system the following inequalities hold:*

$$\varphi_i^1 = \cdots = \varphi_i^p, \quad \beta_i^1 = \cdots = \beta_i^p, \quad i = 1, \ldots, n.$$ (5.2.18)

Proof. Consider a common invariant for all monodromy operators subspace X_l of X.

Let $y_1(\tilde{x}_0), \ldots, y_p(\tilde{x}_0)$ be a Jordan basis for $X|_{\tilde{U}_i^*}$, that is in $y_1, \ldots, y_p$ the matrix $\chi(\sigma_i)$ has a Jordan normal form

$$
G_i = \begin{pmatrix} \rho_i & 1 & & & \\ & \cdot & \cdot & & 0 \\ 0 & & \cdot & \cdot & \\ & & & \cdot & 1 \\ & & & & \rho_i \end{pmatrix}
$$

consisting of only one block. Then the unique filtration of the length p of the space $X|_{\tilde{U}_i^*}$, which is invariant under σ_i^* is the following one:

$$
0 \subset \tilde{X}_1 \subset \cdots \subset \tilde{X}_p = X,
$$

where $\tilde{X}_k$ is generated by $y_1, \ldots, y_k$. Each other filtration can be obtained from this one by uniting of some $\tilde{X}_1, \ldots, \tilde{X}_k$. Thus we get, that $X_l = \tilde{X}_l$ for Levelt's filtration (2.2.9) and therefore $y_1(\tilde{x}_0), \ldots, y_p(\tilde{x}_0)$ is a Levelt's basis in $\tilde{U}_i^*$ for the system. Hence the following inequalities hold:

$$
\varphi_i^1 \geq \cdots \geq \varphi_i^p,
$$

$$
\varphi_i^1 + \cdots + \varphi_i^l \geq \frac{l}{p}(\varphi_i^1 + \cdots + \varphi_i^p), \tag{5.2.19}
$$

$$
(\beta_i^1 + \cdots + \beta_i^l) - \frac{l}{p}(\beta_i^1 + \cdots + \beta_i^p) \geq 0,
$$

where $\varphi_i^k = \varphi_i(y_k)$. (Under the conditions of the proposition, the left hand side of the latter inequality is real).

Note that $\beta_i^1, \ldots, \beta_i^l$ are the exponents of the space X_l at a_i. Summarizing the left hand side of the latter inequality over all $i = 1, \ldots, n$, we get

$$
s_l - \frac{l}{p}s \geq 0, \tag{5.2.20}
$$

where s_l is the sum of exponents of the space X_l, and s is the sum of exponents of system (1.2.1). By Theorem 2.2.2 $s = 0$ and by Lemma 5.2.2 $s_l \leq 0$, therefore by (5.2.20) $s_l = 0$.

From the last equality it follows that (5.2.18) holds. Indeed, if just one of inequalities in (5.2.19) were strict, then by (5.2.20) we would have $s_l - \frac{l}{p}s > 0$, $s_l > 0$, which contradicts the equality $s_l = 0$.

As an immediate corollary of Theorem 5.2.2 and the latter proposition we get the following statement.

Corollary 5.2.2 *Let the monodromy χ of Fuchsian system (1.2.1) be reducible and let each monodromy matrix $\chi(\sigma_i)$ can be transformed to a Jordan normal form, consisting of only one block. Then there exists a constant nonsingular matrix S such that system (5.2.15), obtained from (1.2.1) by changing $z = Sy$ has form (5.2.1).*

5.3   New series of counterexamples

The following statement gives a necessary and sufficient conditions for a representation of some special class to be the representation of some Fuchsian system.

Theorem 5.3.1 *Let χ be a reducible representation with subrepresentation χ_1, and let each monodromy matrix $\chi(\sigma_i), i = 1, \ldots, n$, can be reduced to a Jordan normal form, consisting of only one block. Then Hilbert's 21st problem for the given χ has a positive solution if and only if the following condition holds:*

i) $\gamma_\chi = 0$,

or equivalently

ii) $\gamma_{\chi_1}(0) = \gamma_{\chi_2}(0)$, with $\chi_2 = \chi/\chi_1$.

where $\gamma_\chi(0)$ denotes the weight of the canonical extension G^0 for the bundle E, constructed by χ (see Section 5.1)(cf. [Bo3]).

Proof. Let the problem for a given χ have a positive solution. Then by Corollary 5.2.2 there exists Fuchsian system (1.2.1) with the monodromy χ and with the matrix A of the form

$$A = \begin{pmatrix} A_1 & * \\ 0 & A_2 \end{pmatrix}, \qquad (5.3.1)$$

where A_1, A_2 are the matrices of systems with the corresponding monodromies χ_1, χ_2 respectively. By Proposition 5.2.1 we have that all the valuations $\tilde{\varphi}_i^j$ for all j are equal one together for the both constructed systems. Denote $\tilde{\varphi}_i^j$ by $\tilde{\varphi}_i$.

Let $\tilde{\chi}$ be an arbitrary element of the set $\{\chi, \chi_1, \chi_2\}$. Denote by $\tilde{Y}(\tilde{x})$ a fundamental matrix for the corresponding system of form (1.2.1). By the transformation $Y(\tilde{x}) = \Gamma(x)\tilde{Y}(\tilde{x})$, where

$$\Gamma(x) = \prod_{i=2}^{n}(x - a_i)^{-\tilde{\varphi}_i I}(x - a_1)^{\sum_{i=2}^{n}\tilde{\varphi}_i I} \qquad (5.3.2)$$

our system is transformed to the Fuchsian system with the same singular points and with the valuations

$$\varphi_i^j = 0, \ \varphi_1^j = \varphi_1, \ \text{for all } i = 2, \ldots, n \text{ and for all } j. \qquad (5.3.3)$$

Indeed, the transformation $\Gamma(x)$ has singularities only at $a_1, \ldots, a_n$, since it is equal to $I + o(1)$ at ∞. For a Levelt's fundamental matrix $\tilde{Y}_i(\tilde{x})$ from (2.2.45) we have $\Phi_i = \varphi_i I$, so for the corresponding matrix $Y_i(\tilde{x}) = \Gamma(x)\tilde{Y}_i(\tilde{x})$ we get

$$Y_i(\tilde{x}) = \Gamma(x)\tilde{Y}_i(\tilde{x}) = \Gamma_1(x)V(x)(\tilde{x}_0 - a_i)^{E_i}\chi(\sigma^{-1}), \qquad (5.3.4)$$

where $\tilde{x} = \sigma\tilde{x}_0$, $\tilde{x}_0 \in \tilde{U}_i^*$, $\sigma \in \Delta$ and

$$\Gamma_1(x) = (x - a_1)^{\sum_{i=2}^n \tilde{\varphi}_i I} \prod_{j=2, j\neq i}^n (x - a_j)^{-\tilde{\varphi}_j I}$$

is holomorphically invertible at a_i. Therefore the new system is Fuchsian at $a_2, \ldots, a_n$ and has zero-valuations there. In the similar way one can obtain the second equality in (5.3.3).

As well as in Theorem 5.1.1 we get that the section $W(\tilde{x})$, defined by (5.1.2), is the holomorphic section of the bundle R^λ with $\lambda = (\lambda^1, \ldots, \lambda^n)$, $\lambda^1 = (\varphi_1, \ldots, \varphi_1)$, $\lambda^2 = \cdots = \lambda^n = 0$. In the similar way as in Proposition 5.1.1 and in (5.1.7) we conclude that the bundle G^0 has the splitting type $c_j = \varphi_1$ for all j, therefore $\gamma(0) = 0$.

Let for a given χ condition i) holds. Then by Proposition 5.1.1 the representation χ can be realized as the monodromy representation of some Fuchsian system.

Let for the given χ_1 and χ_2 condition ii) holds. Then the splitting type of the corresponding bundle $G^0_{\chi_1}$ is as follows : $c_1 = \cdots = c_l = c$ (for $\dim \chi_1 = l$) and the splitting type of $G^0_{\chi_2}$ is equal to the next one: $c_1 = \cdots = c_{p-l} = c$.

Indeed, for the first Chern numbers of these bundles we have

$$\sum_{i=1}^n \mathrm{tr}E_i^j = \sum_{i=1}^j c_i = jc, \quad j = l, p - l,$$

where $E_i^l = \frac{1}{2\pi_i}\ln\chi_1(\sigma_i)$, $E_i^{p-l} = \frac{1}{2\pi_i}\ln\chi_2(\sigma_i)$ are the Jordan blocks with the eigenvalues ρ_i, $0 \leq \mathrm{Re}\rho_i < 1$. Thus we get

$$jc = j(\sum_{i=1}^n \rho_i), \quad c = \sum_{i=1}^n \rho_i, \qquad (5.3.5)$$

therefore c is the same for $G^0_{\chi_1}$ and $G^0_{\chi_2}$. Consider the bundles $R^\lambda_{\chi_1}$, $R^\lambda_{\chi_2}$, R^λ_χ, where $\lambda = (\lambda^1, 0\ldots, 0)$, $\lambda^1 = (c, \ldots, c)$. By Proposition 5.1.1 under condition ii) we have that $R^\lambda_{\chi_1}$, $R^\lambda_{\chi_2}$ are holomorphically trivial. Consider holomorphic sections of these bundles and the corresponding fundamental matrices $Y_1(\tilde{x})$ and $Y_2(\tilde{x})$ from (5.1.2) (for the bundles $R^\lambda_{\chi_1}$, $R^\lambda_{\chi_2}$ respectively). Let $W(\tilde{x})$ be a meromorphic section of R^λ_χ such that $W(\tilde{x})$ is holomorphic off a_1 . And let $Y(\tilde{x})$ be the corresponding fundamental matrix from (5.1.2). Then system (1.2.1) with the fundamental matrix $Y(\tilde{x})$ is Fuchsian at $a_2, \ldots, a_n$. By Theorem 5.2.1 we may assume that the matrix $Y(\tilde{x})$ has already form (5.2.5):

$$Y(\tilde{x}) = \begin{pmatrix} T^1 & * \\ 0 & T^2 \end{pmatrix},$$

where T^1, T^2 have the monodromy χ_1 and χ_2 respectively.

Denote by $\Gamma_i(x)$ the following matrices:

$$\Gamma_i(x) = Y_i(\tilde{x})(T^i(\tilde{x}))^{-1}, \ i = 1, 2. \tag{5.3.6}$$

The matrix $\Gamma_i(x)$ is single-valued, since $Y_i(\tilde{x})$ and $T^i(\tilde{x})$ have the same monodromy and $\Gamma_i(x)$ is holomorphically invertible outside of a_1. Indeed, by the construction $Y_i(\tilde{x})$ and $T^i(\tilde{x})$ have the form (5.1.7) (with replacing V by V_i and V^i) with $C = 0$, $\Lambda_1 = 0$ and with the same S_1, $\chi(\sigma^{-1})$. Therefore $\Gamma_i(x) = (V_i^{-1})V^i$ is holomorphically invertible at $a_i, i \neq 1$.

By the matrix

$$\Gamma(x) = \begin{pmatrix} \Gamma_1 & 0 \\ 0 & \Gamma_2 \end{pmatrix}$$

transform our system to the system with the following fundamental matrix: $Y'(\tilde{x}) = \Gamma(x)Y(\tilde{x})$. This system is still Fuchsian at $a_2, \ldots, a_n$ and its fundamental matrix has the next form:

$$Y'(\tilde{x}) = \begin{pmatrix} Y_1(\tilde{x}) & * \\ 0 & Y_2(\tilde{x}) \end{pmatrix} =$$
$$\begin{pmatrix} V_1(x) & W(x) \\ 0 & V_2(x) \end{pmatrix} (x - a_1)^{cI}(\tilde{x} - a_1)^{E_1} \tag{5.3.7}$$

in $\tilde{U}_1^*$, where by construction V_1 and V_2 are holomorphically invertible, $W(x)$ is just meromorphic at a_1.

Let $W(x) = (x - a_1)^{kI}W'(x)$, where $k \leq 0$ and $W'(x)$ is holomorphic at a_1. Then using Lemma 4.1.1, we get that there exists a matrix $\Gamma'(x)$ of form (4.1.5), such that

$$\Gamma'(x)\begin{pmatrix} V_2 \\ W \end{pmatrix} = \begin{pmatrix} V_2 \\ W'' \end{pmatrix}$$

with holomorphic W''. Consider the matrix S such that

$$S\begin{pmatrix} V_2 \\ W'' \end{pmatrix} = \begin{pmatrix} W'' \\ V_2 \end{pmatrix} \text{ and } \tilde{\Gamma} = S\Gamma'S^{-1}. \text{ Then}$$

$$\tilde{\Gamma}\begin{pmatrix} W \\ V_2 \end{pmatrix} = \begin{pmatrix} W'' \\ V_2 \end{pmatrix} \text{ and } \tilde{\Gamma}\begin{pmatrix} V_1 \\ 0 \end{pmatrix} = \begin{pmatrix} V_1 \\ 0 \end{pmatrix}.$$

By $\tilde{\Gamma}$ transform our system with the fundamental matrix $Y'(\tilde{x})$ to the system with the fundamental matrix $Y''(\tilde{x}) = \tilde{\Gamma} Y'(\tilde{x})$. This system is still Fuchsian at $a_2, \ldots, a_n$ and it has a factorization of form (5.3.7) with replacing

$$\begin{pmatrix} V_1(x) & W \\ 0 & V_2 \end{pmatrix} \text{ by } \begin{pmatrix} V_1 & W'' \\ 0 & V_2 \end{pmatrix}.$$

And the latter matrix is already holomorphically invertible at a_1. Thus, our system is Fuchsian at a_1 too.

How to calculate the number $\gamma_\chi(0)$ by the given representation χ? In some cases (for example for $p = 2$) it is possible to do, if the corresponding regular system with the same monodromy is done (see the counterexample of Section 2 and Section 6). In general, it is a difficult problem. But to obtain counterexamples one does not need the exact value of $\gamma_\chi(0)$. It is sufficient to know that $\gamma_\chi(0) > 0$. The following statement provides a simple necessary condition for $\gamma_\chi(0)$ to be equal zero.

Corollary 5.3.1 *Let a representation χ satisfy all conditions of Theorem 5.3.1 (including either i) or ii)). Then for eigenvalues ρ_i of the matrices $E_i = \frac{1}{2\pi i} \ln(\chi(\sigma_i))$ the number*

$$\rho = \sum_{i=1}^{n} \rho_i \tag{5.3.8}$$

is integer.

Proof. The proof immediately follows from (5.3.5), since c is integer.

Example 5.3.1 *Consider the matrices*

$$G_1 = \begin{pmatrix} 1 & 1 & 0 & 0 \\ 0 & 1 & 1 & 0 \\ 0 & 0 & 1 & 1 \\ 0 & 0 & 0 & 1 \end{pmatrix}, \quad G_2 = \begin{pmatrix} 3 & 1 & 1 & -1 \\ -4 & -1 & 1 & 2 \\ 0 & 0 & 3 & 1 \\ 0 & 0 & -4 & -1 \end{pmatrix},$$

$$G_3 = \begin{pmatrix} -1 & 0 & 2 & -1 \\ 4 & -1 & 0 & 1 \\ 0 & 0 & -1 & 0 \\ 0 & 0 & -4 & -1 \end{pmatrix}$$

and any set a_1, a_2, a_3. The representation χ with the singular points a_1, a_2, a_3 and with the monodromy matrices $\chi(\sigma_i) = G_i$, $i = 1, 2, 3$ can not be realized as the monodromy representation of any Fuchsian system.

Proof. Note that $G_1 \cdot G_2 \cdot G_3 \cdot = I$, the matrix G_2 can be transformed to G_1 and the matrix G_3 can be transformed to a Jordan block with the eigenvalue -1. Indeed, for the matrix G_2 we have

$$S_2^{-1} G_2 S_2 = \begin{pmatrix} 1 & 1 & 0 & 0 \\ 0 & 1 & 1 & 0 \\ 0 & 0 & 1 & 1 \\ 0 & 0 & 0 & 1 \end{pmatrix}, S_2 = \frac{1}{3}\begin{pmatrix} 3 & 0 & -1 & 1 \\ -6 & 3 & -1 & 1 \\ 0 & 0 & 3 & 0 \\ 0 & 0 & -6 & 3 \end{pmatrix}$$

and for the matrix G_3 we get

$$S_3^{-1} G_3 S_3 = \begin{pmatrix} -1 & 1 & 0 & 0 \\ 0 & -1 & 1 & 0 \\ 0 & 0 & -1 & 1 \\ 0 & 0 & 0 & -1 \end{pmatrix}, S_3 = \frac{1}{2}\begin{pmatrix} 0 & 8 & 0 & 1 \\ 32 & -8 & 0 & 0 \\ 0 & 0 & 0 & -2 \\ 0 & 0 & -8 & -4 \end{pmatrix}.$$

The representation has a two-dimensional subrepresentation χ_1 with monodromy matrices G_i^1, which are obtained by the intersections of the first two columns and rows of G_i. Thus, the conditions of Theorem 5.3.1 are fulfilled.

By the definition of E_i one has $\rho_1 = \rho_2 = 0$, $\rho_3 = \frac{1}{2\pi i} \ln(-1) = \frac{1}{2}$, therefore the number $\rho = \rho_1 + \rho_2 + \rho_3 = \frac{1}{2}$ is noninteger. By Corollary 5.3.1 we obtain that χ can not be realized as the monodromy representation of any Fuchsian system.

Consider now the next example of a negative solution of Hilbert's 21st problem. From this example it follows that a triangulability of a representation does not ensure a positive answer to the problem.

Example 5.3.2 *Consider any set of points a_1, a_2, a_3, a_4 and the representation, presented by the matrices $G_i = \chi(\sigma_i)$, which equal to the following ones (in order of their appearance):*

$$\begin{pmatrix} 1 & 0 & \underline{1} & 0 & 0 & 1 & 0 \\ 0 & -1 & -1 & 0 & 0 & \underline{1} & 0 \\ 0 & 0 & 1 & \underline{1} & 2 & 2 & 1 \\ 0 & 0 & 0 & 1 & \underline{1} & 1 & 0 \\ 0 & 0 & 0 & 0 & 1 & 1 & \underline{1} \\ 0 & 0 & 0 & 0 & 0 & -1 & 0 \\ 0 & 0 & 0 & 0 & 0 & 0 & 1 \end{pmatrix}, \begin{pmatrix} 1 & 0 & 1 & \underline{1} & -1 & 1 & 0 \\ 0 & -1 & \underline{1} & 1 & -1 & -1 & 0 \\ 0 & 0 & -1 & -1 & \underline{1} & 0 & 0 \\ 0 & 0 & 0 & 1 & 1 & 0 & \underline{1} \\ 0 & 0 & 0 & 0 & -1 & \underline{1} & -1 \\ 0 & 0 & 0 & 0 & 0 & -1 & 0 \\ 0 & 0 & 0 & 0 & 0 & 0 & 1 \end{pmatrix},$$

$$\begin{pmatrix} 1 & 0 & \underline{1} & 0 & -1 & 1 & 1 \\ 0 & -1 & -1 & \underline{1} & 0 & 1 & 0 \\ 0 & 0 & 1 & 1 & -1 & 2 & \underline{1} \\ 0 & 0 & 0 & -1 & \underline{1} & -2 & -1 \\ 0 & 0 & 0 & 0 & -1 & \underline{1} & 1 \\ 0 & 0 & 0 & 0 & 0 & -1 & 0 \\ 0 & 0 & 0 & 0 & 0 & 0 & 1 \end{pmatrix}, \quad \begin{pmatrix} 1 & 0 & 1 & -1 & \underline{1} & 0 & 0 \\ 0 & -1 & 1 & -2 & 1 & 0 & 0 \\ 0 & 0 & -1 & \underline{1} & 0 & 0 & 0 \\ 0 & 0 & 0 & -1 & 1 & \underline{1} & 0 \\ 0 & 0 & 0 & 0 & 1 & -1 & 1 \\ 0 & 0 & 0 & 0 & 0 & -1 & 0 \\ 0 & 0 & 0 & 0 & 0 & 0 & 1 \end{pmatrix}.$$

This representation can not be realized as the monodromy representation of any Fuchsian system.

Proof. Suppose the contrary. Let $Y(\tilde{x})$ be a fundamental matrix of a space X of solutions to Fuchsian system (1.2.1) with the given monodromy. Then the subspaces X_l, $1 \leq l < p$, generated by the first l columns $y_1, \ldots, y_l$ of the matrix $Y(\tilde{x})$ are invariant under the monodromy action. Each matrix G_i can be transformed to a Jordan normal form, consisting of two Jordan blocks with eigenvalues 1 and -1, respectively. Denote vectors of Jordan bases for blocks with eigenvalue 1 by e^i_j and for blocks with eigenvalue -1 by f^i_k. Using only upper-triangular transformations, one can transform each matrix G_i to the form, obtained by the shuffle of these two Jordan blocks. One can obtain this form by replacing by zero all numbers being above the diagonal except for the underlined numbers. The correspoding bases for G_i are as follows:

$$e^1_1, f^1_1, e^1_2, e^1_3, e^1_4, f^1_2, e^1_5, \text{ for } G_1,$$

$$e^2_1, f^2_1, f^2_2, e^2_2, f^2_3, f^2_4, e^2_3, \text{ for } G_2,$$

$$e^3_1, f^3_1, e^3_2, f^3_2, f^3_3, f^3_4, e^3_3, \text{ for } G_3,$$

$$e^4_1, f^4_1, f^4_2, f^4_3, e^4_2, f^4_4, e^4_3, \text{ for } G_4.$$

Note that any vector e^i_1, $i = 1, 2, 3, 4$, generates the subspace X_1, vectors e^i_1, f^i_1 generate the subspace X_2 and so on. Denote by a^i_j the valuation $\varphi_i(e^i_j)$ of the vector e^i_j and by b^i_j the valuation $\varphi_i(f^i_j)$. As it follows from the proof of Proposition 5.2.1 the following inequalities:

$$a^1_1 \geq \cdots \geq a^1_5, \; b^1_1 \geq b^1_2, \tag{5.3.9}$$

$$a^i_1 \geq \cdots \geq a^i_3, \; b^i_1 \geq \cdots \geq b^i_4, \; i = 1, 2, 3$$

hold. It follows from Lemma 5.2.2 that $s_l \leq 0$, where s_l denotes the sum of exponents of X_l. Using Theorem 2.2.2 , (5.3.9) and (5.2.2), one can obtain

$$0 = s_7 = s_6 + a^1_5 + a^2_3 + a^3_3 + a^4_3 \leq s_6 + s_1 \leq 0,$$

therefore $s_6 = 0$ and $s_1 = a_5^1 + a_3^2 + a_3^3 + a_3^4 = 0$. It follows from this equality and (5.3.9) that

$$a_j^i = \cdots = a_5^i = a^i \text{ for some } a^i, i = 1, 2, 3, 4.$$

One can analogously prove that

$$b_1^i = \cdots = b_4^i = b^i, i = 1, 2, 3, 4 \text{ and } s_2 = \sum_{i=1}^{4} b^i + 2 = 0.$$

The condition of Theorem 2.2.2 has in this case the form

$$0 = s_7 = 5a^1 + 3a^2 + 3a^3 + 3a^4 + 2b^1 + 1 + 4b^2 +$$
$$2 + 4b^3 = 2 + 4b^4 + 2 = 2a^1 - 2b^1 - 1, \tag{5.3.10}$$

since $s_1 = s_2 = 0$. But it follows from (5.3.10) that $2a^1 - 2b^1 = 1$ for integers a^1 and b^1, which is impossible. This contradiction means that the given χ can not be realized as the monodromy representation of any Fuchsian system.

5.4 The cases of positive solvability

We present here some sufficient conditions for a reducible representation χ to be the monodromy representation of some Fuchsian system. 70. p.114; 6 row below: replace " for valuations..." by 71. p.114; delete 5 and 4-th rows below.

The first of all we need the following statement.

Lemma 5.4.1 *Suppose that the matrix $G_i = \chi(\sigma_i)$ of an irreducible representation χ has the following form:*

$$G_i = \begin{pmatrix} G_i' & 0 \\ 0 & G_i'' \end{pmatrix}, \tag{5.4.1}$$

where G_i' has the size (t, t), $0 < t < p$. For any integers d_1 and d_2, there exists a Fuchsian system (1.2.1) with the monodromy χ such that its valuations φ_j^k satisfy the following conditions:

i) $\varphi_j^k = d_1$, $j \neq i$, $k = 1, \ldots, p$;

ii) $\varphi_i^k < d_2$, $k = 1, \ldots, t$;

iii) $\varphi_i^m > d_2$, $m = t + 1, \ldots, p$.

Proof. Without loss of generality we may assume that $i = 1$. Let us modify the proof of Theorem 4.2.1 as follows. Consider a regular system (1.2.1) with the given monodromy χ, which is Fuchsian off the singular point a_1 and has valuations φ_j^k, satisfying condition i) of the lemma for $j \neq 1$. (To obtain such a system it is sufficient to consider a meromorphic, holomorphic off a_1 section $W(x)$ of the bundle R^λ, where $\lambda = (\lambda^1, \ldots, \lambda^n)$, $\lambda^j = (d_1, \ldots, d_1)$, $j = 2, \ldots, n$; such a system always exists, see Sections 5.2 and 3.2). Moreover, let the matrix $\chi(\sigma_1)$ have the form (5.4.1) in the basis of the columns of $Y(\tilde{x})$. (Otherwise transform the matrix $Y(\tilde{x})$ to $Y(\tilde{x})S$, where $S^{-1}\chi(\sigma_1)S$ has the form (5.4.1)). Then from the assumption of the lemma it follows that the upper-triangular matrix E_1 in (4.2.1) has the block form

$$E_1 = \mathrm{diag}(E_1', E_2''), \quad \text{where } E_1' = \frac{1}{2\pi i} \ln G_1'.$$

We replace the matrix B in (4.2.1) by $B' = \mathrm{diag}(b_1', \ldots, b_p')$, where all the numbers $b_1', \ldots, b_p'$ satisfy the following conditions:

$$b_1' < d_2 - \frac{(p-1)(n-2)}{2},$$

$$b_p' > d_2 + np + \frac{(p-1)^2(n-2)}{2}, \qquad (5.4.2)$$

$$\sum_{j=1}^{p} b_j' = -pd_1(n-1),$$

$$b_j - b_{j+1} > \frac{p^2(p-1)(n-2)}{2}, \ j = 1, \ldots, t-1, t+1, \ldots, p-1.$$

Let us consider the system with the fundamental matrix $Z(\tilde{x})$, constructed in Theorem 4.2.1. Since $\Phi = D + B'$ has the block form $\Phi = \mathrm{diag}(\Phi', \Phi'')$, where t diagonal elements of the matrix Φ' and $p - t$ elements of the matrix Φ'' are in the nonincreasing order (according to the latter inequality of (5.4.2) and Theorem 4.2.1), the form of E_1 implies that the matrix function

$$(x - a_1)^\Phi E_1 (x - a_1)^{-\Phi}$$

is holomorphic at a_1. Thus, from Corollary 2.2.1 we get that this system is Fuchsian at the point a_1.

From the third condition of (5.4.2), together with the condition

$$\sum_{i=1}^{n} \sum_{j=1}^{p} \beta_i^j = 0$$

for the constructed system (cf. Theorem 2.2.2) and the fact that $0 \leq \mathrm{Re}\rho_j^k < 1$ we get

$$- np < \mathrm{tr}C \leq 0, \tag{5.4.3}$$

where C is from (4.2.2). Indeed,

$$0 = \sum_{i=1}^{n} \sum_{j=1}^{p} \beta_i^j = \sum_{i=2}^{n} \sum_{j=1}^{p} \beta_i^j + \sum_{j=1}^{p} \beta_1^j =$$

$$= pd_1(n-1) + \sum_{i=2}^{n} \sum_{j=1}^{p} \rho_i^j + \mathrm{tr}(D + B') + \sum_{j=1}^{p} \rho_1^j =$$

$$= \sum_{i=1}^{n} \sum_{j=1}^{p} \rho_i^j + \mathrm{tr}C + pd_1(n-1) + \mathrm{tr}B',$$

since $\mathrm{tr}C = \mathrm{tr}D$ by (4.2.3). The inequalities (5.4.3) follow now from the inequality

$$0 \leq \sum_{i=1}^{n} \sum_{j=1}^{p} \rho_i^j < pn,$$

(which in turn follows from $0 \leq \mathrm{Re}\rho_i^j < 1$) and from the third equality in (5.4.2):

$$\mathrm{tr}B' = \sum_{j=1}^{p} b_j' = -pd_1(n-1).$$

From (4.2.6) and (5.4.3) we get

$$c_1 p - \mathrm{tr}C \leq \frac{(n-2)p(p-1)}{2},$$

$$c_1 p \leq \frac{(n-2)p(p-1)}{2} + \mathrm{tr}C \leq \frac{(n-2)p(p-1)}{2},$$

$$c_1 \leq \frac{(n-2)(p-1)}{2}. \tag{5.4.4}$$

From (5.4.4) and the condition of nonincreasing for $c_1, \ldots, c_p$ it follows that

$$c_1 + \cdots + c_{p-1} \leq c_1(p-1) \leq \frac{(n-2)(p-1)^2}{2}.$$

Therefore from the left hand side inequality in (5.4.3) we get

$$c_p > -np - \frac{(n-2)(p-1)^2}{2}.$$

(5.4.5)

Since for all $j \leq t$ and for all l by (5.4.4) , (5.4.5) and by the first condition in (5.4.2) we have

$$b'_j + c_l < b'_1 + c'_1 < d_2,$$

we obtain that the first t diagonal elements of Φ do not exceed d_2. In the similar way for all $j > t$ and for all l by the second condition in (5.4.2) we get

$$b'_j + c_l > b'_p + c_p > d_2,$$

therefore the last $p - t$ diagonal elements of Φ are less than d_2. So we get conditions ii) and iii) for the fundamental matrix $Z(\tilde{x})$.

Theorem 5.4.1 *Let all matrices $\chi(\sigma_i)$ of the monodromy representation χ can be simultaneously transformed to the following form:*

$$G_j = \begin{pmatrix} G_j^1 & * \\ 0 & G_j^2 \end{pmatrix},$$

(5.4.6)

where the size of each G_j^1 is (l, l). Let the collection of the matrices $G_1^i, \ldots, G_n^i$ define the representation χ_i, $i = 1, 2$ and let the representation χ_2 can be realized as the monodromy representation of some Fuchsian system. If χ_1 is irreducible and for some i the matrix G_i has the form (5.4.1) with $t \leq l$, then the monodromy χ also can be realized as the monodromy representation of some Fuchsian system.

Proof. Without loss of generality we can assume that $i = 1$. The proof of the theorem is similar to the proof of the sufficiency of condition ii) for Theorem 5.3.1.

Consider a fundamental matrix $Y_2(\tilde{x})$ of a Fuchsian system with the monodromy χ_2, such that $\chi_2(\sigma_j) = G_j^2$ in the basis of the columns of $Y(\tilde{x})$. Let

$$\max_{i=2,\ldots,n; j=1,\ldots,p-l} \varphi_i^j = d_1, \quad \max_{j=1,\ldots,p-l} \varphi_1^j = d_2$$

(5.4.7)

for the valuations of the system. Consider the corresponding fundamental matrix $Y_1(\tilde{x})$ of the Fuchsian system , constructed in the previous lemma, with the monodromy χ_1 and with d_1 and d_2 from (5.4.7). Let $\chi_1(\sigma_j) = G_j^1$ in the basis of the columns of $Y_1(\tilde{x})$.

As in Theorem 5.3.1, let $Y(\tilde{x})$ be a fundamental matrix of a regular system with the monodromy χ with Fuchsian singular points $a_2, \ldots, a_n$. In addition, let $Y(\tilde{x})$ have the form

$$Y(\tilde{x}) = \begin{pmatrix} T^1 & * \\ 0 & T^2 \end{pmatrix}, \tag{5.4.8}$$

where T^1 and T^2 have the monodromy matrices G_j^1, G_j^2, $j = 1, \ldots, n$, respectively.

Consider the matrices Γ_i from the formula above (5.3.7) and the new fundamental matrix $Y'(\tilde{x}) = \Gamma(x)Y(\tilde{x})$ with

$$\Gamma = \begin{pmatrix} \Gamma_1 & 0 \\ 0 & \Gamma_2 \end{pmatrix}.$$

For the corresponding Levelt's matrix $Y_i'(\tilde{x}) = Y_i(\tilde{x})S_i$, $i \neq 1$ we have factorization (2.2.45) with the holomorphic matrix $V(x)$ and with $\Phi_i = \mathrm{diag}(d_1, \ldots, d_1, \varphi_i^1, \ldots, \varphi_i^{p-l})$ From (5.4.7) it follows that the diagonal elements of Φ_i form a nonincreasing sequence. Therefore by Corollary 2.2.1 we get that system (1.2.1) with the fundamental matrix $Y'(\tilde{x})$ is Fuchsian at a_i for $i > 1$.

In $\tilde{U}_1^*$ the matrix $Y_1(\tilde{x})$ has the form

$$Y_1(\tilde{x}) = \begin{pmatrix} V_1(x) & W(x) \\ 0 & V_2(x) \end{pmatrix} (x - a_1)^{\Phi_1} (\tilde{x} - a_1)^{E_1}. \tag{5.4.9}$$

By the second condition in (5.4.7) and condition iii) of Lemma 5.4.1 we obtain that for diagonal elements φ_1^j of Φ_1 the following inequalities hold:

$$\varphi_1^1 \geq \cdots \geq \varphi_1^t; \quad \varphi_1^{t+1} \geq \cdots \geq \varphi_1^p.$$

Since the matrix E_1 has the same block form as G_1 from (5.4.1), we get that the matrix

$$L(x) = \Phi_1 + (x - a_1)^{\Phi_1} E_1 (x - a_1)^{-\Phi_1}$$

is holomorphic at a_1. Indeed, $L(x)$ also has a block form and a holomorphy of such blocks follows from the formula below (2.2.27).

In the similar way as in Theorem 5.3.1, by $\tilde{\Gamma}(x)$ transform our system to the system with the fundamental matrix $\tilde{Y}(x) = \tilde{\Gamma}Y'(\tilde{x})$, which has a holomorphic factor

$$\begin{pmatrix} V_1 & W'' \\ 0 & V_2 \end{pmatrix}$$

in the corresponding factorization (2.2.45). By Corollary 2.2.1 the constructed system is Fuchsian at a_1.

Remark 5.4.1 *The statement of the theorem does not depend on a disposition of the irreducible representation* χ_1 *in (5.4.6). If we assume that the collection* $G_1^2, \ldots, G_n^2$ *defines the irreducible representation of the dimension* l *and the size of* G_i' *from (5.4.1) is greater than* $p - l - 1$, *then we shall obtain the same result for* χ *as in the theorem (under assumption that* χ_1 *can be realized as the monodromy representation of some Fuchsian system).*

The proof is just the same as in the theorem with one difference. We must replace (5.4.7) by

$$\min_{i=2,\ldots,n; j=1,\ldots,p-l} \varphi_i^j = d_1, \quad \min_{j=1,\ldots,p-l} \varphi_1^j = d_2. \tag{5.4.10}$$

Corollary 5.4.1 *Let all matrices* $\chi(\sigma_i)$ *of a monodromy representation* χ *can be simultaneously transformed to the form (5.4.6), where the size of each* G_j^1 *is* (l, l) . *Let for some* i

$$G_i = \begin{pmatrix} G_i^1 & 0 \\ 0 & G_i^2 \end{pmatrix}.$$

If the both representations χ_1, χ_2 *with matrices* $\chi_i(\sigma_j) = G_j^i$, $i = 1, 2$ *can be realized as the monodromy representations of some Fuchsian systems, then the same is also true for the representation* χ.

Proof. The proof is a simplified version of the proof of Theorem 5.4.1. One should make the following changes: drop all conditions related to d_2; set $t = l = \dim \chi_1$; choose a Fuchsian system with the monodromy χ_1 so that its valuations φ_j^k for $j \neq i$ satisfy the condition $\varphi_j^k > d_1$, $k = 1, \ldots, l$. The last condition is easily satisfied by means of the transformation

$$\tilde{Y}_1(\tilde{x}) = \left(\prod_{j \neq i} (x - a_j)^d \right) (x - a_i)^{-d(n-1)} Y_1(\tilde{x})$$

with $d = d_1 + 1 - \min_{j \neq i; k=1,\ldots,l} \varphi_j^k$. (Note that the matrix of this transformation is holomorphic at ∞).

The following statement is the direct corollary of Theorem 5.4.1 and Corollary 5.4.1.

Theorem 5.4.2 *Let all matrices* $\chi(\sigma_i)$ *can be simultaneously transformed to form (5.2.14), where each collection* $G_1^j, \ldots, G_n^j$ *forms the irreducible representation* χ_j. *If for each* j *there exists* i *such that the matrix* G_i *with help of an upper-triangular matrix* S_i^j *can be transformed to the form* $G_i' = S_i^j G_i (S_i^j)^{-1}$, *where*

$$
G_i' = \begin{pmatrix}
(G_i^1)' & & * & \Big| & & & 0 \\
& \ddots & & \Big| & & & \\
0 & & (G_i^j)' & \Big| & & & \\
- & - & - & \Big| & - & & - \\
& & & \Big| & (G_i^{j+1})' & & * \\
& 0 & & \Big| & & \ddots & \\
& & & \Big| & 0 & & (G_i^k)'
\end{pmatrix}, \qquad (5.4.11)
$$

then the representation χ can be realized as the monodromy representation of some Fuchsian system.

Below, in Section 5.5 we present some other new sufficient conditions for the positive solvability of Hilbert's 21st problem .

5.5   On regular systems

The negative solution of Hilbert's 21st problem means that, as distinct from a local situation, the class of Fuchsian systems and the class of systems with regular singular points are not meromorphically equivalent globally throughout the Riemann sphere. In other words, there are systems (1.2.1) with regular singular points $a_1, \ldots, a_n$ that cannot be reduced to a Fuchsian system by any change

$$
z = \Gamma(x)y \qquad (5.5.1)
$$

of the unknown vector function y by means of the matrix $\Gamma(x)$ meromorphic on $\bar{\mathbb{C}}$, which is holomorphically invertible outside of the points $a_1, \ldots, a_n$.

But each regular system occurs to be a subsystem (quotient system) of some regular system, which is already meromorphically equivalent to a Fuchsian one.

This statement is equivalent to the following one.

Theorem 5.5.1 *Any representation χ is a subrepresentation (quotient representation) of some representation for which Hilbert's 21st problem has a positive solution.*

To prove this theorem we need some new special condition for a positive solvability of the problem.

Proposition 5.5.1 *Let the monodromy matrices $\chi(\sigma_i)$ of the representation χ can be simultaneously transformed to the following form*

$$
G_l = \begin{pmatrix} \underline{G_l^1} & & & * \\ & |\underline{G_l^2} & & \\ & & \cdot & \\ 0 & & & |\underline{G_l^m} \end{pmatrix}, \quad l = 1, \ldots, n. \tag{5.5.2}
$$

where for all $j = 1, \ldots, m$ the collection of the matrices $G_1^j, \ldots, G_n^j$ is irreducible and the size of G_i^j is (k_j, k_j). If all subrepresentations of χ are indecomposable and for some i the matrix G_i has a block form

$$
G_i = \begin{pmatrix} G_i' & 0 \\ 0 & G_i'' \end{pmatrix}, \tag{5.5.3}
$$

where the size l of the matrix G_i' does not exceed k_1, then Hilbert's 21st problem for the representation χ has a positive solution.

Proof. Denote by χ_j the subrepresentation, formed by intersections of the first $k_1 + \cdots + k_j$ rows and columns of the matrices G_l. Without loss of generality, we can take it that the matrix G_1 has the required form (5.5.3) and that it is upper-triangular.

Let us consider a system with regular singular points $a_1, \ldots, a_n$, which is Fuchsian at the points $a_2, \ldots, a_n$, and a fundamental matrix $Y(\tilde{x})$ for the space of solutions of this system of form (5.5.2) (with replacing G_l^j by the corresponding Y^j), in the basis of whose columns the matrix $\chi(\sigma_i)$ has form (5.5.3). (The existence of a system with such a matrix $Y(\tilde{x})$ is proved in Corollary 5.2.1). Suppose that

$$
Y(\tilde{x}) = V_1(x)(\tilde{x} - a_1)^{E_1} \tag{5.5.4}
$$

in $\tilde{U}_1^*$, then the matrix $V_1(x)$, meromorphic at the point a_1 also has form (5.5.2) with replacing G_1^k by V_1^k. Denote by r the order of the zero (the pole with the sign minus) of the function $\det V_1(x)$ at the point a_1 and by r_1 the order of the zero (the pole with the sign minus) of the function $\frac{\det V_1(x)}{\det V_1^1(x)}$.

Let $d_1, d_2, d_3, b_1, \ldots, b_p$ be integers for which the following inequalities are satisfied:

$$
d_1 > b_1 > \ldots > b_l,
$$

$$
b_{l+1} > \ldots > b_p > d_2,
$$

$$
b_j - b_{j+1} > d_3, \quad j \neq l,
$$

$$r - ld_1 - (p - l)b_{l+1} > 0,$$

$$r_1 - (p - k_1)d_2 + \frac{(n - 2)(p - k_1)(p - k_1 - 1)}{2} < 0. \qquad (5.5.5)$$

We represent the matrix $Y(\tilde{x})$ from (5.5.4) in form (4.2.1)

$$Y(\tilde{x}) = V_1'(x)(x - a_1)^B (\tilde{x} - a_1)^{E_1}, \qquad (5.5.6)$$

where $B = \mathrm{diag}(b_1, \ldots, b_p)$, b_j from (5.5.5). Let us consider the corresponding matrix $\Gamma_1(x)$ and factorization (4.2.2) for the matrix $V_1'(x)$:

$$\Gamma_1(x)V_1'(x) = (x - a_1)^C V(x) \qquad (5.5.7)$$

with a holomorphically invertible at a_1 matrix $V(x)$ and $C = \mathrm{diag}(c_1, \ldots, c_p)$, $c_1 \geq \cdots \geq c_p$.

Lemma 5.5.1 *Inequality (4.2.6) is satisfied under the conditions of Proposition 5.5.1 for the elements of the matrix C from (5.5.7).*

Proof. We shall prove the lemma by induction with respect to the number m of blocks in (5.5.2). For $m = 1$ the lemma follows from Proposition 4.2.1. Suppose that the statement has been proved for $m = t$. Let us prove it for $m = t + 1$.

It follows from (5.5.6) and (5.5.7) that $\sum_{i=1}^p c_i = r - \sum_{i=1}^p b_i$, and therefore, by virtue of (5.5.5),

$$c_1 \geq \frac{1}{p} \sum_{i=1}^p c_i \geq \frac{1}{p} \left(r - \sum_{i=1}^l b_i - \sum_{i=l+1}^p b_i \right) \geq$$

$$\geq \frac{1}{p}(r - ld_1 - (p - l)b_{l+1}) > 0. \qquad (5.5.8)$$

Let us consider the first row $(y_1, \ldots, y_p)$ of the matrix $\Gamma_1(x)Y(\tilde{x})$.

Lemma 5.5.2 *The element $y_1(\tilde{x})$ of the first row of the matrix $\Gamma_1(x)Y(\tilde{x})$ is not identically zero.*

Proof. We suppose that $y_1(\tilde{x}) \equiv 0$. Then $y_2(\tilde{x}) \equiv \ldots \equiv y_{k_1}(\tilde{x}) \equiv 0$ as well by virtue of the irreducibility of the subrepresentation χ_1 (see Lemma 4.2.1). Since the matrix $Y(\tilde{x})$ has form (5.5.2), the last identity means that the first k_1 elements in the first row γ of the matrix $\Gamma_1(x)$ are zero. Since all the elements of Γ_1 are holomorphic outside of the point a_1, it follows that $\gamma = (0, \ldots, 0, \gamma_1, \ldots, \gamma_{p-k_1})$, where there are two possibilities: a) $\gamma_j = s \neq 0$, $s \in \mathbb{C}$ for a certain j, $1 \leq j \leq p - k_1$; b) all nonzero γ_j have poles at 0.

Consider the case a). Let us consider a matrix

$$\hat{\Gamma}(x) = \begin{pmatrix} \gamma_1 \dots & s & \dots \gamma_{p-k_1} \\ I^{j-1} & 0 & 0 \\ 0 & 0 & I^{p-k_1-j} \end{pmatrix},$$

(where I^l is the identity matrix of the size l) and a matrix $Y'(\tilde{x})$ we obtain from $Y(\tilde{x})$ by deleting the first k_1 rows and columns. We denote by V_1'', B' and E_1' the matrices obtained after we delete the first k_1 rows and columns of the matrices V_1', B and E_1 respectively. We find from (5.5.7) that

$$\hat{\Gamma}(x)V_1''(x) = \hat{V}(x), \tag{5.5.9}$$

where the first row $\hat{v}(x)$ of the matrix $\hat{V}(x)$ has the form $\hat{v}(x) = (x - a_1)^{c_1}v(x)$, $v(x)$ is a row vector holomorphic at the point a_1.

In $\tilde{U}_1^*$ the matrix $Y'(\tilde{x})$ has the factorization of form (5.5.6):

$$Y'(\tilde{x}) = \hat{V}_1'(x)(x - a_1)^{B'}(\tilde{x} - a_1)^{E_1'}. \tag{5.5.10}$$

We consider a respective matrix $\Gamma'(x)$ and the factorization of form (5.5.7) with matrices C' and $V'(x)$:

$$\Gamma'(x) \cdot \hat{V}_1'(x) = (x - a_1)^{C'}V'(x). \tag{5.5.11}$$

According to the induction assumption

$$\sum_{i=1}^{p-k_1}(c_1' - c_i') \le \frac{(n-2)(p-k_1)(p-k_1-1)}{2}. \tag{5.5.12}$$

On the other hand, by virtue of the second inequality in (5.5.5)

$$\sum_{i=1}^{p-k_1} c_i' = r_1 - (b_{k_1+1} + \dots + b_p) < r_1 - (p-k_1)d_2, \tag{5.5.13}$$

where r_1 is the order of the zero of the function $\frac{\det V_1}{\det V_1^1} = \det \hat{V}_1'$ at the point a_1. Adding inequalities (5.5.12) and (5.5.13) together and taking the last inequality in (5.5.5) into account, we have $(p - k_1)c_1' < 0$, whence

$$c_1' < 0. \tag{5.5.14}$$

Let us consider a matrix

$$Z(x) = \hat{\Gamma}(x)Y'(\tilde{x})(\Gamma'(x)Y'(\tilde{x}))^{-1}.$$

On one hand, the matrix $Z(x)$ is meromorphic and holomorphically invertible outside off the point a_1. On the other hand, for the first row $z_1(x)$ of this matrix in the neighborhood of the point a_1 from (5.5.10), (5.5.9) and (5.5.11) we obtain

$$z_1(x) = (x - a_1)^{c_1 - c_1'} v(x)(V'(x))^{-1},$$

where $v(x)$ is a row holomorphic at the point a_1. From the fact that $v(x)$ is holomorphic and from inequalities (5.5.8) and (5.5.14) we get $z_1(a_1) = 0$. Since the row $z_1(x)$ is holomorphic throughout the Riemann sphere outside of a_1, it follows, according to Liouville's theorem, that $z_1(x) \equiv 0$, and this contradicts the holomorphic invertibility of the matrix $Z(x)$ outside of the point a_1. Thus, $y_1(\tilde{x})$ is not identically equal to zero.

Consider the case b). From Lemma 4.1.1 it follows that there exists a meromorphic matrix W of the form

$$\begin{pmatrix} I^l & 0 \\ 0 & W' \end{pmatrix},$$

which is holomorphically invertible outside of 0 and such that the first row γ' of $\Gamma_1 W$ has the form, required in the item a). (More precisely this fact follows from the lemma, applied to Γ_1^t, where t means transposition. In this case one must take $-c_{k_1+i}$ equal to the order of pole of γ_i at zero, $l = 1$, $Y = x^{c_{k_1+j}} \gamma_j$ with maximal c_{k_1+j}. The procedure of Lemma 4.1.1 decreases the pole of γ_j. Iterating this process, as a result we obtain a matrix W_0 of the required form and such that $W_0 \Gamma_1^t$ has the first column which contains either only holomorphic elements at zero or only zero elements excepting one, say γ_m'. But in the latter case γ_m' cannot have a pole at zero too, because in opposite case this function would have a zero at some x' and it would be $\det(W_0(x') \Gamma_1^t(x') = 0$, which contradicts the equality $\det \Gamma_1(x) = \text{const} \neq 0$. Let $W = W_0^t$. This completes the proof). The proof of the lemma in the case b) repeats now the proof of the item a) with replacing γ by γ', $Y(\tilde{x})$ by $W^{-1} Y(\tilde{x})$, and V_1 by $W^{-1} V_1$.

As it follows from Remark 4.2.1 all we need for completeness of the proof of Lemma 5.5.1 is the linear independence of the functions $y_1, \ldots, y_p$. The corresponding proof is presented in Lemma 5.5.3. The only thing that we else have to verify is the holomorphy of $(x - a_1)^B E_1 (x - a_1)^{-B}$. But since the matrix $E_1 = \frac{1}{2\pi i} \ln \chi(\sigma_1)$ is of the block form (5.5.3), we have to verify this condition for each block separately. And we can carry out the verification, with due account of the first two inequalities in (5.5.5), in the same way as we did in Proposition 4.2.1.

Lemma 5.5.3 *Suppose that the first element $y_1(x)$ of a row $y_1, \ldots, y_p$ of the matrix $Y(\tilde{x})$ is nonzero. If any subrepresentation of representation χ of the monodromy of system (1.2.1) is indecomposable then the functions $y_1, \ldots, y_p$ are linearly independent.*

Proof. Denote by X the space of solutions for the system. Let X_l be the subspace of X, generated by the first $k_1 + \cdots + k_l$ columns of Y. Let us assume that the statement of the lemma is false. Then there is an upper-triangular nonsingular matrix S such that the elements with indices $j \in J$ of the row vector $(x_1, \ldots, x_p) = (y_1, \ldots, y_p)S$ are linearly independent and the elements with indices $j \in \{1, \ldots, p\} \backslash J = J^o$ are identically zero. Let us consider a linear subspace X^0 of the vector functions that is generated by the columns t'_j of the matrix YS, $\quad j \in J^o$.

For arbitrary monodromy operator σ^* we have

$$\sigma^*(t'_j) = \sum_{l=1}^{p} \lambda_l t'_l, \tag{5.5.15}$$

where $t'_j \in X^0$. We get from (5.5.15) that

$$0 \equiv (x_j \circ \sigma) = \sum_{l=1}^{p} \lambda_l x_l = \sum_{l \in J} \lambda_l x_l,$$

for the row of the matrix $Y(\tilde{x})S$ being considered, whence, by virtue of the linear independence of the elements x_l, $l \in J$, we have $\lambda_l = 0$, $l \in J$. Therefore, the right hand side of (5.5.15) is a linear combination of the elements of the space X^0. In other words, X^0 is an invariant subspace for all monodromy operators.

By the hypothesis, X^0 does not contain the first column of the matrix $Y(\tilde{x})$, and therefore $X_1 \not\subset X^0$, and this means that $X_1 \cap X^0 = \emptyset$ since otherwise χ_1 would be reducible. Suppose that s is the first number for which $X_s \cap X^0 \neq \emptyset$, and then $X_{s-1} \cap X^0 = \emptyset$ and, since the quotient representation χ_s / χ_{s-1} is irreducible, we find that all columns of the matrix YS with numbers $k_1 + \ldots k_{s-1} + 1, \ldots, k_1 + k_s$ belong to X^0, i.e. $X_s \cap X^0$ is an invariant subspace for χ_s and $\chi_s = \chi_{s-1} \oplus \chi_s / \chi_{s-1}$, and this contradicts the assumption concerning the irreducibility of the subrepresentation χ_s of the representation χ. Lemma 5.5.3 is proved.

Let us return to the proof of Proposition 5.5.1. According to Lemma 5.5.1, there is a matrix $\Gamma(x)$, meromorphic and holomorphically invertible outside of the point a_1, such that

$$\Gamma(x)Y(\tilde{x}) = (x - a_1)^C V(x)(x - a_1)^B (\tilde{x} - a_1)^{E_1}, \tag{5.5.16}$$

where $C = \mathrm{diag}(c_1, \ldots, c_p)$, $\quad c_1 \geq \ldots \geq c_p$, the numbers c_i satisfy inequality (4.2.6), $V(x)$ is holomorphically invertible, B is a diagonal integral matrix whose elements satisfy the first three inequalities in (5.5.5). We choose a number d_3 in (5.5.5) such that

$$d_3 > \frac{(n-2)p(p-1)}{2}.$$

The final part of the proof of Proposition 5.5.1 is exactly the same as of Theorem 4.2.1. There is only one difference. In our case the corresponding matrix $D+B$ from (4.2.4) has a block form. These blocks correspond the blocks of E_1 and diagonal elements of each block form a nonincreasing sequence. Therefore, the matrix

$$L = D + B + (x - a_1)^{D+B} E_1 (x - a_1)^{-D-B}$$

is holomorphic at a_1 and we also can use Corollary 2.2.1.

Proposition 5.5.2 *Suppose that all quotient representations of a representation χ are indecomposable. If the monodromy matrices of this representation can be simultaneously reduced to form (5.5.2) so that the matrix $\chi(\sigma_i)$ has form (5.5.3) for a certain i, where the size of the matrix G_i'' does not exceed k_m, then Hilbert's 21st problem for the representation χ has a positive solution.*

Proof. As well, as in Proposition 5.5.1, we again suppose that $i = 1$, and again consider a system with regular singular points $a_1, \ldots, a_n$, Fuchsian at the points $a_2, \ldots, a_n$, and a fundamental matrix $Y(\tilde{x})$ of the space of solutions of this system of form (5.5.2) in the basis of whose columns the matrix $\chi(\sigma_1)$ is upper-triangular and has form (5.5.3). Let us consider factorization (5.5.4) for this matrix.

Let d_1, d_2, κ, $b_1, \ldots, b_p$ be integers for which the equalities

$$\begin{aligned}
b_i &= d_1 - \kappa(i - 1), \quad i = 1, \ldots, l, \\
b_j &= d_2 + \kappa(p - j), \quad j = l + 1, \ldots, p, \\
\kappa &> 0,
\end{aligned} \qquad (5.5.17)$$

hold true, where l is the size of the matrix G_1', $\quad l \geq p - k_m$.

Let us represent the matrix $Y(\tilde{x}))$ from (5.5.4) in form (5.5.6), where $B = \text{diag}(b_1, \ldots, b_p)$, b_i from (5.5.17). We consider the corresponding matrix $\Gamma_1(x)$ and factorization (5.5.7) for the matrix $V_1'(x)$. We denote by $(y_1, \ldots, y_p)$ the first row of the matrix ΓY and, as we did in Lemma 5.5.3, consider an upper-triangular nonsingular matrix S such that the elements with indices $j \in J$ of the row vector $(x_1, \ldots, x_p) = (y_1, \ldots, y_p) S$ are linearly independent and the elements with indices $j \in J^0 = \{1, \ldots, p\} \backslash J$ are identically zero. As in Lemma 5.5.3, we can use the irreducibility of the representation $\chi_s \backslash \chi_{s-1}$ to prove that $J = J_{m_1} \cup \ldots \cup J_{m_t}$, where J_{m_k} is the set of numbers of columns of some block G_i^s in (5.5.2). The monodromy matrices of the fundamental matrix $\Gamma(x) Y(\tilde{x})$ are the same as of $Y(\tilde{x})$, and therefore we find from the form of the row $x = (x_1, \ldots, x_p)$ that the elements with numbers (k, s) of the matrices $G_1' = S^{-1} G_i S$, where $k \in J$, $s \in J^0$ are zero. Indeed, for $s \in J^0$ we get

$$0 \equiv (x_s \circ \sigma_i) = x_s(\sigma_i^{-1}\tilde{x}) = \sum_{k \in J} g_{ks} x_k,$$

whence, by virtue of the linear independence of x_k, $k \in J$, we get $g_{ks} = 0$, $k \in J$, $s \in J^0$, for elements of G'_i.

We denote by U'_i, B', V', H'_i, Φ'_i the matrices obtained as a result of the deleting of the rows and columns with the numbers from J^0 of the matrices $V'_i (x - a_1)^B S (x - a_1)^{-B}$, B, V, G'_i, Φ_i respectively.

Note that the numbers of columns of the last block of G_i^m from (5.5.2) are not contained in J^0, since otherwise the quotient representation χ/χ_{s-1}, where G_i^s is the last block, the numbers of whose columns enter into J would be decomposable. Therefore, the matrix S has the form

$$S = \begin{pmatrix} \hat{S} & 0 \\ 0 & I^{k_m} \end{pmatrix}, \tag{5.5.18}$$

where $\hat{S}$ is an upper-triangular matrix. From this and from (5.5.17) we find that the matrix $(x - a_1)^B S (x - a_1)^{-B}$ is holomorphic at a_1 and $\det((x - a_1)^B S (x - a_1)^{-B}) = \det S \neq 0$. Consequently, V' is a holomorphically invertible matrix.

The row $(x_{i_1}, \ldots, x_{i_q})$, $i_j \in J$, $q = |J|$, is analytic outside of the points $a_1, \ldots, a_n$ and has monodromy $H'_1, \ldots, H'_n$ and factorizations of form (4.2.10) with replacing y_j by x_j, Φ_i by Φ'_i, B by B', E_i by $E'_i = \frac{1}{2\pi i} \ln H'_i$, $S_i = I$.

Applying the procedure used in Proposition 4.2.1 to this row, we get an equality of form (4.2.20):

$$qc_1 + \sum_{j \in J} b_j + \sum_{i=2}^{n} \sum_{j \in J} \varphi_i^j +$$

$$+ \sum_{i=2}^{n} \sum_{j \in J} \rho_i^j + d = \frac{(n-2)q(q-1)}{2}, \tag{5.5.19}$$

where ρ_i^j, $\rho_i^j \geq 0$ are diagonal elements of the matrix E_i, $d \geq 0$. We denote by $\hat{d}$ the sum of all positive numbers from φ_i^j, $i = 1, \ldots, n$, $j = 1, \ldots, p$; then $\sum_{i=2}^{n} \sum_{j \in J} \varphi_i^j \leq \hat{d}$, and it follows from (5.5.19) that

$$pc_1 \leq \frac{(n-2)p(p-1)}{2} - \frac{p}{q} \sum_{j \in J} b_j. \tag{5.5.20}$$

Relation (4.2.9) holds true for the form $\omega = dY \cdot Y^{-1}$. Subtracting it from (5.5.20), we get

$$\sum_{i=1}^{p}(c_1 - c_i) \leq K + \sum_{i=1}^{p} b_i - \frac{p}{q}\sum_{j\in J} b_j, \tag{5.5.21}$$

where K is a constant independent of B.

Let us evaluate the difference $D = \sum_{i=1}^{p} b_i - \frac{p}{q}\sum_{j\in J} b_j$. From the fact that the numbers from $p - l + 1$ to p are contained in J and from (5.5.17) we have

$$D = \sum_{i=1}^{l} b_i - \frac{p}{q}\sum_{j\in\{1,\ldots,l\}\cap J} b_j - \frac{p-q}{q}\sum_{j=l+1}^{p} b_j \leq$$

$$\leq \sum_{i=1}^{l} b_i - \frac{p}{q}\sum_{j=l-q_1+1}^{l} b_j - \frac{p-q}{q}\sum_{j=l+1}^{p} b_j,$$

where $q_1 = |\{1,\ldots,l\}\cap J|$. Using (5.5.17) again, we obtain after simple calculations

$$D \leq \frac{(p-q)(p-l)}{q}(d_1 - d_2 + \kappa(1-p)) + \kappa\frac{p(p-q)}{2}. \tag{5.5.22}$$

Indeed, by (5.5.17) we have

$$\sum_{i=1}^{l} b_i = ld_1 - \kappa\frac{l(l-1)}{2}; \quad \sum_{j=l-q_1+1}^{l} b_j = q_1 d_1 - \kappa\frac{(2l - q_1 - 1)q_1}{2},$$

$$\sum_{j=l+1}^{p} b_j = (p-l)d_2 + \kappa\frac{(p-l)(p-l-1)}{2},$$

therefore,

$$D \leq ld_1 - \kappa\frac{l(l-1)}{2} - \frac{pq_1}{q}d_1 + \frac{\kappa p}{q}\frac{(2l - q_1 - 1)q_1}{2} -$$

$$- \frac{(p-q)(p-l)}{q}d_2 - \frac{\kappa}{q}\frac{(p-q)(p-l)(p-l-1)}{2} = M + N,$$

where

$$M = ld_1 - \frac{pq_1}{q}d_1 - \frac{(p-q)(p-l)}{q}d_2,$$

$$N = -\kappa\frac{l(l-1)}{2} + \frac{\kappa p}{q}\frac{(2l - q_1 - 1)q_1}{2} - \frac{\kappa}{q}\frac{(p-q)(p-l)(p-l-1)}{2}.$$

Since $q = p - l + q_1$, it follows that

$$M = \frac{(p-q)(p-l)}{q}(d_1 - d_2)$$

and

$$N = \frac{\kappa}{2q}[-ql(l-1) + p(l+(p-q)-1)(l-(p-q)) - (p-q)(p-l)(p-l-1)] =$$

$$= \frac{\kappa}{2q}[-ql(l-1) + p(l^2-(p-q)^2 - l + (p-q)) - (p-q)(p-l)(p-l-1)] =$$

$$= \frac{\kappa}{2q}[(l-1)(-ql+pl) + (p-q)(-p(p-q)+p-(p-l)^2+p-l)] =$$

$$= \frac{\kappa(p-q)}{2q}[l^2-l-p^2+pq+p-p^2+2pl-l^2+p-l] = \frac{\kappa(p-q)}{2q}[2(p-l)(1-p)+pq] =$$

$$= \frac{(p-q)(p-l)\kappa(1-p)}{q} + \frac{\kappa(p-q)p}{2}.$$

Adding M and N, we get (5.5.22). Suppose now that

$$\kappa > p \cdot |K|.$$

We choose d_2 and d_1 such that the number D in (5.5.22) does not exceed 0. (If $q = p$, then $D = 0$ and if $q < p$, then $\frac{(p-q)(p-l)}{q} > \frac{p-l}{p} > 0$ and we must choose $d_1 < d_2$ such that

$$d_2 - d_1 > \frac{\kappa p^2(p-q)}{2(p-l)}.)$$

Now (5.5.21) yields

$$c_1 - c_p \le \sum_{i=1}^{p}(c_1 - c_i) \le |K| \le \frac{\kappa}{p}. \tag{5.5.23}$$

The last part of the proof of the proposition completely repeats the corresponding part of Theorem 4.2.1. We have only to verify that the elements b'_i of the matrix $D + B$ satisfy the following conditions

$$b'_1 \ge \cdots \ge b'_l, \quad b'_{l+1} \ge \cdots \ge b'_p, \tag{5.5.24}$$

to use the block form (5.5.3) of the matrix E'_1, and to use Corollary 2.2.1. But the inequalities (5.5.24) follow from (5.5.17) and (5.5.23).

Proof of Theorem 5.5.1.   It is sufficient to show that every representation χ is a subrepresentation (quotient representation) of some representation which satisfies the conditions of Proposition 5.5.2 (Proposition 5.5.1).

Let us prove the second statement (the first statement can be proved by analogy).

Suppose that representation χ of form (5.5.2) has decomposable subrepresentations $\chi_{s_1}, \ldots, \chi_{s_r}, \quad s_1 < \ldots < s_r, \quad r > 1$. Let $s = s_2$.

We take it that the matrices G_2^2 and G_3^s have been reduced to the Jordan normal form by means of the matrices R^2 and R^s (to make this assumption, it is sufficient to pass to the matrices $\hat{G}_i = R^{-1} G_i R$, where R has form (5.5.2) (with replacing G_i^j by R^j) with $R^k = I, \quad k \neq 2, s$). Let $t_2 = g_{11}^2, \quad t_3 = g_{11}^3$, where $G_3^s = (g_{ij}^s)$, $G_2^2 = (g_{ij}^2)$.

Let us consider representation χ of size $p + 1$ with matrices

$$G_1' = \begin{pmatrix} t_1 & 0 \\ 0 & G_1 \end{pmatrix}, \qquad G_2' = \begin{pmatrix} t_2 & 1\,0\ldots0 \\ 0 & G_2 \end{pmatrix},$$

$$G_3' = \begin{pmatrix} t_3 & t_{11}\ldots t_{1p} \\ 0 & G_3 \end{pmatrix}, G_i' = \begin{pmatrix} 1 & 0 \\ 0 & G_i \end{pmatrix}, \quad i > 3,$$

where $t_{1j} = -\frac{1}{t_2}\hat{g}_{1j}, \quad t_1 = \frac{1}{t_2 t_3}, \quad G_3 = (\hat{g}_{ij})$.

This representation already has $r - 1$ decomposable subrepresentations and the matrices G_i' have form (5.5.3) which satisfy the conditions of Proposition 5.5.1. Repeating the described procedure, after $r - 1$ steps we shall get the representation $\hat{\chi}$ satisfying the conditions of Proposition 5.5.1 whose quotient representation is the original representation χ.

If representation χ does not have decomposable subrepresentations (or $r = 1$) and the number m of blocks in (5.5.2) exceeds unity, then, in the general case, we have to carry out once the procedure described above all the same (say, for $s = 2$, s_1) in order to obtain the matrix G_i' of form (5.5.3). It suffices to reduce only the matrix G_2^s to the Jordan normal form.

Corollary 5.5.1 *The size of the representation we have constructed does not exceed the number $p + m - 1$, where m is the number of irreducible blocks in factorization (5.5.2) for the monodromy matrices of the original representation χ.*

From Theorem 5.5.2 and Corollary 5.5.1 we obtain the following statement.

Theorem 5.5.2 *Any system (1.2.1) with regular singular points is a subsystem (quotient system) of some system , which is meromorphically equivalent to a Fuchsian system with the same singular points. The size of the latter system does not exceed the number $p + m - 1$, where p is the size of the original system, and m is the number of blocks in decomposition (5.5.2) for the monodromy representation χ of the system (1.2.1).*

Example 5.5.1 *Let us consider the following system with regular singular points $0, -1, 1, \frac{1}{2}$:*

$$
\frac{dy}{dx} = \left(\left(\begin{array}{cccc} 0 & 0 & 0 & 0 \\ 0 & 0 & 1 & 0 \\ 0 & 0 & x & 0 \\ 0 & 0 & 0 & -x \end{array} \right) \frac{1}{x^2} + \frac{1}{6} \left(\begin{array}{cccc} 0 & 6 & 0 & 0 \\ 0 & 0 & 6 & 0 \\ 0 & 0 & -1 & 1 \\ 0 & 0 & -1 & 1 \end{array} \right) \frac{1}{x+1} + \right.
$$

$$
\left. + \frac{1}{2} \left(\begin{array}{cccc} 0 & 0 & 0 & 0 \\ 0 & 0 & 0 & 2 \\ 0 & 0 & -1 & -1 \\ 0 & 0 & 1 & 1 \end{array} \right) \frac{1}{x-1} + \frac{1}{3} \left(\begin{array}{cccc} 0 & -3 & 0 & 0 \\ 0 & 0 & -3 & -3 \\ 0 & 0 & -1 & 1 \\ 0 & 0 & -1 & 1 \end{array} \right) \frac{1}{x-\frac{1}{2}} \right) y
$$

(5.5.25)

Its quotient system resulting after the deletion of the first rows and columns of the matrices of coefficients in (5.5.25) is not meromorphically equivalent to any Fuchsian system, since for the monodromy representation of this quotient system Hilbert's 21st problem has a negative solution (see Section 2.4). As to system (5.5.25) itself, we can use a change $z = \Gamma_2(x)\,\Gamma_1(x)y$ with matrices

$$
\Gamma_1(x) = \left(\begin{array}{cccc} 1 & 0 & 0 & 0 \\ -\frac{1}{3x} & 1 & 0 & 0 \\ 0 & 0 & 1 & 0 \\ 0 & 0 & 0 & 1 \end{array} \right), \quad \Gamma_2(x) = \left(\begin{array}{cccc} 1 & 0 & 0 & 0 \\ 0 & 1 & -\frac{1}{x} & 0 \\ 0 & 0 & 1 & 0 \\ 0 & 0 & 0 & 1 \end{array} \right)
$$

to reduce it to the Fuchsian system

$$
\frac{dz}{dx} = \left(\left(\begin{array}{cccc} 1 & 0 & 3 & 0 \\ -\frac{1}{3} & -1 & -2 & 0 \\ 0 & 0 & 1 & 0 \\ 0 & 0 & 0 & -1 \end{array} \right) \frac{1}{x} + \frac{1}{6} \left(\begin{array}{cccc} -2 & 6 & -6 & 0 \\ \frac{2}{3} & 2 & 3 & 1 \\ 0 & 0 & -1 & 1 \\ 0 & 0 & -1 & 1 \end{array} \right) \frac{1}{x+1} + \right.
$$

$$
\left. + \frac{1}{2} \left(\begin{array}{cccc} 0 & 0 & 0 & 0 \\ 0 & 0 & 1 & 3 \\ 0 & 0 & -1 & -1 \\ 0 & 0 & 1 & 1 \end{array} \right) \frac{1}{x-1} + \frac{1}{3} \left(\begin{array}{cccc} -2 & -3 & -6 & 0 \\ \frac{2}{3} & 2 & 3 & -5 \\ 0 & 0 & -1 & 1 \\ 0 & 0 & -1 & 1 \end{array} \right) \frac{1}{x-\frac{1}{2}} \right) z.
$$

(5.5.26)

Since any system of two equations with n regular singular points and any system of three equations with three regular singular points are meromorphically equivalent to Fuchsian systems (see Dekkers's result in Section 3.4 and Chapter 6), Example 5.5.1 is minimal in this sense (with respect to the number of equations in the system and the number of singular points).

Remark 5.5.1 *System (5.5.25) is also the first (with respect to the number of equations and singular points) example of a Fuchsian system which cannot be reduced by any meromorphic change to a Fuchsian system with the same singular points the matrices of whose coefficients have the same character of reducibility (5.5.2) as its monodromy matrices.*

Indeed, the matrices of system (5.5.26) have the form (5.5.2) with $m = 2$, $k_1 = 2$, $k_2 = 2$ and as it was shown, they could not be reduced to form (5.5.2) with $m = 3$ blocks of the sizes $k_1 = 1$, $k_2 = 1$, $k_3 = 2$ respectively. But the monodromy matrices of (5.5.26) have exactly the latter character of reducibility (it follows from the fact that this character of reducibility have the matrices of system (5.5.25)).

The first example of such a type was presented in Proposition 2.2 in [Bo4] (for the monodromy representation of dimension $p = 7$).

Example 5.5.2 *Let us consider matrices*

$$G_1' = \begin{pmatrix} 1 & 1 & 0 & 0 & 0 \\ 0 & 1 & 1 & 0 & 0 \\ 0 & 0 & 1 & 1 & 1 \\ 0 & 0 & 0 & 1 & 1 \\ 0 & 0 & 0 & 0 & 1 \end{pmatrix}, \qquad G_2' = \begin{pmatrix} 3 & 1 & 1 & -1 & 0 \\ -4 & -1 & 1 & 2 & 0 \\ 0 & 0 & 3 & 1 & 0 \\ 0 & 0 & -4 & -1 & -1 \\ 0 & 0 & 0 & 0 & 1 \end{pmatrix},$$

$$G_3' = \begin{pmatrix} -1 & 0 & 2 & -1 & 0 \\ 4 & -1 & 0 & 1 & 0 \\ 0 & 0 & -1 & 0 & 0 \\ 0 & 0 & 0 & 1 & 0 \\ 0 & 0 & 0 & 0 & 1 \end{pmatrix} \qquad (5.5.27)$$

and an arbitrary collection of points a_1, a_2, a_3. Representation χ with matrices G_1', G_2', G_3' satisfies the conditions of Proposition 5.5.2 (G_3' has form (5.5.3)), and therefore Hilbert's 21st problem has a positive solution for it. At the same time, the subrepresentation of size 4 of this representation with matrices G_j, formed by intersections of the first four rows and columns of the former matrices G_1', G_2', G_3' cannot be the monodromy representation of any Fuchsian system (see Example 5.3.1).

As it was proved by Plemelj (see Section 3.2) each representation χ can be realized as the monodromy representation of some regular system (1.2.1), which is Fuchsian at all singular points $a_1, \ldots, a_n$ except one of them (say a_1). How to estimate the order of the pole of system (1.2.1) at a_1? The simple estimate is presented in the following proposition.

Proposition 5.5.3 *For arbitrary representation χ there exists a regular system (1.2.1) Fuchsian off the point a_1 such that the order of the pole of (1.2.1) at a_1 does not exceed the number*

$$d = \frac{(n-2)p(p-2)}{2} + pn + 1.$$

Proof. Let χ be of form (5.5.2). Consider a regular system (1.2.1) Fuchsian off a_1 with a fundamental matrix $Y(\tilde{x})$ of the form (5.5.2) (see Corollary 5.2.1):

$$Y(\tilde{x}) = \begin{pmatrix} \underline{Y^1} & & * \\ & |\underline{Y^2} & \\ & & \ddots \\ 0 & & |\underline{Y^m} \end{pmatrix} \tag{5.5.28}$$

with zero valuations at the points $a_2, \ldots, a_n$.

By Lemma 4.1.3 it follows that for suitable $\Gamma^i(x)$ we have in $\tilde{U}_1^*$

$$\Gamma^i(x)Y^i(\tilde{x}) = V^i(x)(x - a_1)^{D_i}(\tilde{x} - a_1)^{E_1^i}, \tag{5.5.29}$$

where by Proposition 4.2.1:

$$\sum_{j=1}^{k_i}(d_i^0 - d_i^j) \leq \frac{(n-2)k_i(k_i - 1)}{2}. \tag{5.5.30}$$

Here $d_i^0 = \max_{j=1,\ldots,k_i}(d_i^j)$, $D_i = \operatorname{diag}(d_i^1, \ldots, d_i^{k_i})$, k_i is the size of Y^i. As well as in (5.4.3) of Lemma 5.4.1 we have

$$-k_i n < \operatorname{tr} D_i, \tag{5.5.31}$$

$$\operatorname{tr} D_i \leq 0.$$

Adding (5.5.30) with the last inequality, we get

$$k_i d_i^0 \leq \frac{(n-2)k_i(k_i - 1)}{2},$$

$$d_i^0 \leq \frac{(n-2)(k_i-1)}{2}. \tag{5.5.32}$$

Transform our system to the system with the fundamental matrix $Y'(\tilde{x}) = \Gamma(x)Y(\tilde{x})$, where

$$Y(\tilde{x}) = \begin{pmatrix} \underline{\Gamma^1} & & & 0 \\ & |\underline{\Gamma^2} & & \\ & & \cdot & \\ 0 & & & |\underline{\Gamma^m} \end{pmatrix}.$$

Then from (5.5.29) we obtain in $\tilde{U}_1^*$

$$Y'(\tilde{x}) = \begin{pmatrix} \underline{V^1}| & \underline{W^{12}}| & \underline{W^{13}}| & 0 \\ 0 & |\underline{V^2} & |\underline{W^{23}} & \\ & & \cdot & \\ 0 & & & |\underline{V^m} \end{pmatrix} (x-a_1)^D (x-a_1)^{E_1}, \tag{5.5.33}$$

where $V^1, \ldots, V^m$ are holomorphically invertible at a_1, W^{ij} are just meromorphic there, $D = \mathrm{diag}(D_1, \ldots, D_m)$, and

$$E_1 = \begin{pmatrix} E_1^1 & & * \\ & \ddots & \\ 0 & & E_1^m \end{pmatrix}.$$

In the same way as in Theorem 5.3.1 we can transform our system by $\Gamma_{12}(x)$ to the system with holomorphic matrix W^{12} (see the part of the proof of Theorem 5.3.1 below (5.4.8)). By the next step we can get the holomorphy of W^{13} and W^{23} and so on. As a result we obtain the system with a fundamental matrix of the form (5.5.33) with holomorphic W^{ij} for all i,j.

It follows from (2.2.29) that the order of the pole of the constructed system is equal to $r+1$, where r is the order of the pole of

$(x-a_1)^D E_1 (x-a_1)^{-D}$, where $D = \mathrm{diag}(D_1, \ldots, D_m)$, $E_1 = \mathrm{diag}(E_1^1, \ldots, E_1^m)$.

B ut $r+1$ does not exceed the number

$$d = \max_{i,j} d_i^j - \min_{i,j} d_i^j + 1 \leq \sum_{k,l} (\max_{i,j} d_i^j - d_k^l) + 1 =$$

$$= p \max_{i,j} d_i^j - \sum_{i=1}^m \mathrm{tr} D_i + 1. \tag{5.5.34}$$

From (5.5.32) it follows that

$$p \cdot \max_{i,j} d_i^j \leq \frac{(n-2)p(p-2)}{2}, \tag{5.5.35}$$

since $\max k_i \leq p - 1$.

Summarizing (5.5.31) over all i, adding the result with (5.5.34), and taking into account (5.5.35) we get

$$d \leq \frac{(n-2)p(p-2)}{2} + pn + 1.$$

The exact estimate for the case $p = 3$ is presented in [Bo2] (see also Chapter 6).

5.6   On codimension of "nonfuchsian" representations

How to estimate the codimension of representations, which can not be realized as the monodromy representations of Fuchsian systems? (We call them "nonfuchsian" in this section).

Let $a_1, \ldots, a_n$ be fixed. The moduli space $\mathcal{R}$ of all irreducible representations χ of dimension p with the singular points $a_1, \ldots, a_n$ depends on

$$N_r = p^2(n-2) + 1 \tag{5.6.1}$$

parameters. Indeed, from the equality $G_1 \cdots G_n = I$ it follows that a representation χ is determined by $n-1$ matrices $G_1, \ldots, G_{n-1}$; they give $p^2(n-1)$ parameters. Since the representation χ is determined up to conjugations $G_i' = S^{-1}G_iS, i = 1, \ldots, n-1$ and since the equalities $S_1^{-1}G_iS_1 = S_2^{-1}G_iS_2, i = 1, \ldots, n$ imply $S_1S_2^{-1} = \lambda I$, $\lambda \in \mathbb{C}^*$ (Schur's lemma), then we obtain that the number of parameters must be decreased by $p^2 - 1$. Thus,

$$N_r = p^2(n-1) - (p^2 - 1) = p^2(n-2) + 1.$$

Reducible representations lie in the boundary of $\mathcal{R}$ and form a stratification of the boundary. Nonfuchsian representations lie among these strata.

Consider a representation χ of form (5.5.2) with $m = 2$

$$G_i = \begin{pmatrix} G_i^1 & * \\ 0 & G_i^2 \end{pmatrix} \tag{5.6.2}$$

and consider the subrepresentation χ_1, defined by the collection $G_1^1, \ldots, G_n^1$. Let $\dim \chi_1 = l$ and $l \leq \frac{p}{2}$. Each matrix G_i has $l(p-l)$ zeros, therefore we obtain

$$N_1 = l(p - l)(n - 1) \tag{5.6.3}$$

additional conditions for such a χ.

If Hilbert's 21st problem for a representation χ of form (5.6.2) has a negative solution, then from Theorem 5.4.1 we get the following statement (which was observed by V.Kostov)

Lemma 5.6.1 *Let $k_1, \ldots, k_t$ be sizes of Jordan blocks $J'_1, \ldots, J'_t$ in a Jordan normal form for G^1_i. If the representation χ can not be realized as the monodromy representation of any Fuchsian system, then for each J'_q there exists the corresponding Jordan block J''_q of the size m_q in a Jordan normal form for G^2_i, which is adjoint to J'_q and does not adjoint to other blocks.*

Proof. With help of a transformation

$$S = \begin{pmatrix} S_1 & * \\ 0 & S_2 \end{pmatrix}$$

transform each matrix G^1_i, G^2_i to a Jordan normal form, such that each Jordan block J''_q of G^2_i would be adjoint at most to one block J'_q of G'_i. If we obtain some block in G^2_i, which is not adjoint to blocks of G^1_i, then there exists a matrix

$$S' = \begin{pmatrix} S'_1 & * \\ 0 & S'_2 \end{pmatrix},$$

such that $(S')^{-1} G_i S'$ has form (5.5.3) and we are under assumptions of Theorem 5.4.1. Thus χ can be realized as the monodromy representation of some Fuchsian system, but this contradicts the assumption of the lemma. Hence, all blocks of G''_i are adjoint to the corresponding blocks of G'_i. In the similar way one can prove that for each block of G'_i there exists a block of G'''_i, which is adjoint to the first one. The lemma is proved.

By Lemma 5.6.1 we have

$$k_1 + \cdots + k_t = l,$$

$$m_1 + \cdots + m_t = p - l.$$

The existence of a Jordan block of the size $k_j + m_j$ in G_i admits $k_j + m_j - 1$ additional conditions (coincidence of $k_j + m_j - 1$ eigenvalues of G_i). Thus, we have N_2 additional conditions:

$$N_2 = ((k_1 + m_1 - 1) + \cdots + (k_t + m_t - 1))n - t = (p - t)n - t, \tag{5.6.4}$$

where $t \leq l$. This formula needs the following explanation. Eigenvalues of the corresponding blocks in $G_1, \ldots, G_{n-1}$ do not determine uniquely an eigenvalue of the corresponding block in G_n, but by the equality $G_1 \cdots G_n = I$ they determine it up to $\sqrt[r]{1}$, where $r = k_q + m_q$ for the corresponding q. For example, if $l = 1$ and $t = 1$ for all $i = 1, \ldots, n$, then each G_i has only one eigenvalue $\mu_i = g_{11}^i$, where $G_i = (g_{11}^i)$. From $G_1 \cdots G_n = I$ we have that $g_{11}^n = (g_{11}^1 \cdots g_{11}^{n-1})^{-1}$ is determined uniquely by $g_{11}^1, \ldots, g_{11}^{n-1}$. But for eigenvalue μ_n of the G_n^2 (if we know that it has only one eigenvalue) we obtain only $\mu_n = g_{11}^n \sqrt[r-1]{1}$. Indeed,

$$(\mu_n)^{p-1} = \det G_n^2 = \left(\sum_{i=1}^{n-1} \det G_i^2 \right)^{-1} = (g_{11}^n)^{p-1}$$

and we get the previous formula. Since the number of possible values for eigenvalue μ_n is finite, we have that this does not influence on the codimension. Therefore, for G_n we obtain only $p - 2$ (not $p - 1$) additional conditions, which are the conditions for G_n^2 to have only one eigenvalue.

Thus, in general case (when $t > 1$) we have to subtract t in the right hand side of (5.6.4).

If $l = 1$, then by Theorem 5.3.1 we have to add one more condition, that is $\gamma(\chi/\chi_1) > 0$. (For $l = 1$ the number ρ from (5.3.8) is integer, therefore this condition can not be satisfied automatically). In general, this condition can be expressed in terms of the corresponding Fuchsian system with the monodromy χ/χ_1. It is algebraic equation on coefficients of the system (see Chapter 6) and therefore with help of the holomorphic map from the set of Fuchsian systems into $\bar{\mathcal{R}}$ (which assign to each system its monodromy) we obtain one analytic condition.

And at last we must add some number r conditions, which arise from the fact that the collection $G_1, \ldots, G_n$ is determined up to conjugations. Thus, for codimension d we get (since $l \geq t$):

$$d \geq N_1 + N_2 + r + \delta_l^1 = l(p - l)(n - 1) + (p - t)n - t + r + \delta_l^1 \geq$$

$$(l+1)(p-l)(n-1) + (p-l) - l + r + \delta_l^1 > 2(p-1)(n-1) + p - 1 + r, \quad (5.6.5)$$

since the quadratic polynomial $(l + 1)(p - l)(n - 1) - 2l$ in l, regarded in $[1, \frac{p}{2}]$, has a minimum at $l = 1$. (Here δ_l^1 is the Kronecker index). Thus,

$$d \geq 2(p - 1)(n - 1) + p - 1. \quad (5.6.6)$$

Let us show that for nonfuchsian representations of form (5.5.2) with $m = 2$ and $l = 1$ this estimate is exact. For this purpose it is sufficient to prove that $r = 0$ in (5.6.5), since $d = 2(p - 1)(n - 1) + p - 1 + r$ in this case.

Since a centralizer S of one Jordan block has the form

$$S = \begin{pmatrix} \alpha & \beta_1 & \cdots & \cdot & \beta_{p-1} \\ & \cdot & \cdot & \cdots & \cdot \\ & & \cdot & \cdot & \cdot \\ & & & \cdot & \beta_1 \\ 0 & & & & \alpha \end{pmatrix} \qquad (5.6.7)$$

(see [Ga]), then from

$$SG_k S^{-1} = G_k, \quad k = 1, \ldots, n$$

we obtain that S has the form (5.6.7) with some $\beta_1, \ldots, \beta_{p-1}$. From the irreducibility of the collection $G_1^2, \ldots, G_n^2$ we get, that

$$S = \begin{pmatrix} \alpha & * \\ 0 & \alpha I^{p-1} \end{pmatrix},$$

therefore $\beta_1 = \cdots = \beta_{p-2} = 0$. Since there exists i such that $g_{pt}^i \neq 0$, $1 < t < p$ for the elements g_{pt}^i of G_i (it follows from irreducibility of χ/χ_1), we obtain from the equality $SG_i = G_i S$, that $\beta_{p-1} = 0$. Thus, $S = \alpha I$, as in irreducible case, and we get $r = 0$ in (5.6.6).

We investigated here only the case of a reducibility of the type $m = 2$ in (5.5.2). But it is obvious, that for sufficiently large p and n the codimension of nonfuchsian representations with $m > 2$ is larger than (5.6.6), since the number of zeros in (5.5.2) increases as $\kappa \cdot p \cdot m$ for some κ. So we get the following statement.

Theorem 5.6.1 *For sufficiently large p and n the exact codimension for the main stratum of nonfuchsian representations in the moduli space of all representations is equal to*

$$d = 2(p-1)(n-1) + p - 1 = (p-1)(2n-1).$$

Questions concerning a stratification of nonfuchsian representations and their codimension in $(\mathrm{GL}(p, \mathbb{C}))^{n-1}$ are also considered in [Ko1], [Ko2].

6 The case $p = 3$

For the case $p = 3$ there is a complete description of all representations, which cannot be realized as the monodromy representations of Fuchsian systems.

6.1 The complete answer for $p = 3$

It occurs that Theorem 5.3.1 gives all counterexamples to Hilbert's 21st problem for $p = 3$.

Theorem 6.1.1 *Hilbert's 21st problem for a representation χ of dimension $p = 3$ has a negative solution if and only if the following three conditions hold:*

a) the representation χ is reducible;

b) each matrix $\chi(\sigma_i)$ can be reduced to a Jordan normal form, consisting of only one block;

c) the corresponding two-dimensional subrepresentation or quotient representation χ_2 is irreducible [1] *and the weight $\gamma_{\chi_2}(0)$ of the canonical extension G^0 for χ_2 is greater than zero.*

Proof. By Theorem 4.2.1, Theorem 5.3.1 and Plemelj's result (cf. Section 3.2) it follows that all we need is to prove a positive solvability of the problem for reducible representations, possessing the following property: for a given χ there exists i such that the Jordan normal form of the matrix $\chi(\sigma_i)$ consists of two Jordan blocks. For such a representation there are two possibilities :

i) χ is upper-triangular;

ii) χ is not upper-triangular.

Let us consider the case i). In this case all $\chi(\sigma_i)$ can be simultaneously transformed to the form

---

[1] the condition of irreducibility follows from others, but it is convenient to have it for concrete verifications

$$G_j = \begin{pmatrix} \mu_j^1 & & * \\ 0 & \mu_j^2 & \\ 0 & 0 & \mu_j^3 \end{pmatrix}, \; j = 1, \ldots, n. \tag{6.1.1}$$

Let for some $j = i$ two of the numbers μ_j^k are different. If

$$\mu_i^1 = \mu_i^2 \neq \mu_i^3 \text{ or } \mu_i^2 = \mu_i^3 \neq \mu_i^1,$$

then with help of an upper-triangular matrix S we can transform the matrix G_j to the form

$$G_i' = S^{-1}G_iS = \begin{pmatrix} G_i^1 & 0 \\ 0 & G_i^2 \end{pmatrix}, \tag{6.1.2}$$

where the size l of G_i^1 equals two in the first case ($\mu_i^1 = \mu_i^2 \neq \mu_i^3$) and $l = 1$ in the second one. In the both cases we are under assumptions of Corollary 5.1, since due to Dekkers each two-dimensional representation can be realized as the monodromy representation of some Fuchsian system (cf. Section 3.4). Therefore the representation χ also can be realized by a Fuchsian system.

Let now for some i $\mu_i^1 = \mu_i^3 \neq \mu_i^2$. We may consider only the case $\mu_j^1 = \mu_j^3$ for all $j = 1, \ldots, n$, since otherwise we are under condition of the previous situation (for some j). By an upper-triangular matrix S transform G_i to the matrix

$$G_i' = S^{-1}G_iS = \begin{pmatrix} \mu_i^1 & 0 & 1 \\ 0 & \mu_i^2 & 0 \\ 0 & 0 & \mu_i^1 \end{pmatrix}. \tag{6.1.3}$$

Consider a fundamental matrix $Y(\tilde{x})$ of the form

$$Y(\tilde{x}) = \begin{pmatrix} y^1 & & * \\ 0 & y^2 & \\ 0 & 0 & y^3 \end{pmatrix} \tag{6.1.4}$$

to regular system (1.2.1) with the given monodromy χ such that the system is Fuchsian outside of a_1 and has zero-valuations there (the existence of such a system is proved in Corollary 5.2.1). Moreover, let G_i has form (6.1.3) in the basis of the columns of $Y(\tilde{x})$ in $\tilde{U}_i^*$. Then for y^1 and y^3 in each $\tilde{U}_t^*$ we have

$$y^j(\tilde{x}) = v_t^j(x)(\tilde{x} - a_t)^{\rho_t}, \; t \neq i,$$

$$y^j(\tilde{x}) = v_i^j(x)(x - a_i)^{\varphi_i^j}(\tilde{x} - a_t)^{\rho_i},$$

where $j = 1, 3$, v_t^j is holomorphically invertible for all j, t, $\exp(2\pi i\rho_t) = \mu_t^1 = \mu_t^3$.

By the theorem on the sum of residues, applied to the form $d \ln y^j(\tilde{x})$ we get

$$\sum_{j=1}^{n} \rho_j + \varphi_i^1 = \sum_{j=1}^{n} \rho_j + \varphi_i^2 = 0,$$

therefore $\varphi_i^1 = \varphi_i^2$. Thus, in $\tilde{U}_i^*$ for the matrix $Y(\tilde{x})$ we have the following factorization (2.2.25):

$$Y(\tilde{x}) = \begin{pmatrix} v^1 & w^{12} & w^{13} \\ 0 & v^2 & w^{23} \\ 0 & 0 & v^3 \end{pmatrix} (x-a_i) \begin{pmatrix} \varphi_i^1 & 0 & 0 \\ 0 & \varphi_i^2 & 0 \\ 0 & 0 & \varphi_i^1 \end{pmatrix}_{(\tilde{x}-a_i)} \begin{pmatrix} \rho_i^1 & 0 & \varepsilon \\ 0 & \rho_i^2 & 0 \\ 0 & 0 & \rho_i^1 \end{pmatrix},$$

$$(6.1.5)$$

since G_i has form (6.1.3) in the basis of the columns of $Y(\tilde{x})$. Here v^1, v^2, v^3 are holomorphically invertible at a_i; w^{12}, w^{13}, w^{23} are meromorphic there.

With help of the procedure of Lemma 4.1.1 we can obtain the holomorphy of w^{kl}. Indeed, if

$$w^{12} = \frac{c_1}{(x - a_1)^{r_1}} + w',$$

then we can decrease the order of the pole with help of the transformation $Y'(\tilde{x}) = \Gamma_1(x)Y(\tilde{x})$, where

$$\Gamma_1(x) = \begin{pmatrix} 1 & -\frac{c_1}{(x-a_1)^{r_1} v^2(a_1)} & 0 \\ 0 & 1 & 0 \\ 0 & 0 & 1 \end{pmatrix}.$$

This transformation does not change the element v^1. After finite number of such transformations we shall obtain the holomorphy of $\tilde{w}^{12}$. Then, we shall do the similar procedure with w^{23}, w^{13}, using v^3 and a transformation of the form

$$\Gamma(x) = \begin{pmatrix} 1 & 0 & Q_1(x) \\ 0 & 1 & Q_2(x) \\ 0 & 0 & 1 \end{pmatrix},$$

which do not change elements v^1, w^{12}, v^2.

Thus, there exists $\Gamma(x)$ such that $\tilde{Y}(\tilde{x}) = \Gamma(x)Y(\tilde{x})$ has form (6.1.5) with holomorphic w^{kl} for all k, l. It means that the corresponding system is Fuchsian at a_1, since

$$(x - a_i) \begin{pmatrix} \varphi_i^1 & 0 & 0 \\ 0 & \varphi_i^2 & 0 \\ 0 & 0 & \varphi_i^1 \end{pmatrix} \begin{pmatrix} \rho_i^1 & 0 & \varepsilon \\ 0 & \rho_i^2 & 0 \\ 0 & 0 & \rho_i^1 \end{pmatrix} (x - a_i)^{-\begin{pmatrix} \varphi_i^1 & 0 & 0 \\ 0 & \varphi_i^2 & 0 \\ 0 & 0 & \varphi_i^1 \end{pmatrix}}$$

is holomorphic at a_i and we are under assumptions of Corollary 2.2.1.

The case $\mu_i^1 = \mu_i^2 = \mu_i^3$, $i = 1, \ldots, n$ is a simplified version of the previous one (because $\varphi_i^1 = \varphi_i^2$ in (6.1.5)).

Let us consider the case ii). In this case all $\chi(\sigma_j)$ can be simultaneously transformed to the form

$$G_j = \begin{pmatrix} G_j^1 & * \\ 0 & G_j^2 \end{pmatrix},$$ (6.1.6)

where both the collections of the matrices $\{G_j^1\}$ and $\{G_j^2\}$ form irreducible representations. Consider the matrix G_i, whose Jordan normal form consists of two blocks. Then the one-dimensional block of this matrix belongs to some G_i^j, $j = 1, 2$ and is not ajoint to other block. This means that the condition of Lemma 5.6.1 is not fulfilled, therefore this representation can be realized as the monodromy representation of some Fuchsian system. (In the case i) the situation differed from this one, because a two-dimensional collection of the corresponding matrices was reducible and we could not apply Lemma 5.6.1).

6.2 Fuchsian weight of a representation

It is not difficult problem to verify conditions a), b) of Theorem 6.1.1 for a given χ. But how to calculate the number $\gamma_{\chi_2}(0)$? In this section we present a partial answer on this question.

Consider a Fuchsian system of two equations

$$\frac{dy}{dx} = \left(\sum_{i=1}^{n} \frac{B^i}{x - a_i} \right) y$$ (6.2.1)

with a monodromy χ and valuations $\{\varphi_i^j\}$, $i = 1, \ldots, n$, $j = 1, 2$. Recall (cf. Section 2.3), that the number

$$\gamma_B = \sum_{i=1}^{n} |\varphi_i^1 - \varphi_i^2|$$

is called *the Fuchsian weight of the system* and the number

$$\gamma(\chi) = \min_B \gamma_B$$

(over all Fuchsian systems (6.2.1) with the given χ) is called *the Fuchsian weight of the representation χ*.

The main purpose of this section is to prove the following statement

Theorem 6.2.1 *The weight of the canonical extension G^0 for a two dimensional representation χ coincides with its Fuchsian weight:*

$$\gamma_\chi(0) = \gamma(\chi).$$

To prove the theorem the first of all we need the following proposition.

Proposition 6.2.1 *For each Fuchsian system (6.2.1) with Fuchsian weight γ_B there exists a Fuchsian system $\frac{dy}{dx} = \tilde{B}y$ with the same monodromy χ and there exists an index $l, 1 \le l \le n$, such that the following conditions hold:*

i) $\tilde{\varphi}_i^1 = \tilde{\varphi}_i^2 = 0$, $i = 1, \ldots, n$, $i \ne l$,

ii) $\gamma_{\tilde{B}} = \tilde{\varphi}_l^1 - \tilde{\varphi}_l^2 \le \gamma_B$

for valuations $\tilde{\varphi}_i^j$ and for Fuchsian weight $\gamma_{\tilde{B}}$ of the new system.

Proof. For each i we consider factorization (2.2.25) for system (6.2.1):

$$Y_i(\tilde{x}) = V_i(x)(x - a_i)^{\Phi_i}(\tilde{x} - a_i)^{E_i}, \quad \tilde{x} \in \tilde{U}_i^*, \tag{6.2.2}$$

where $V_i(x) = (v_{mt}^i)$, $\Phi_i = \mathrm{diag}(\varphi_i^1, \varphi_i^2)$,

$$E_i = \begin{pmatrix} \rho_i^1 & \varepsilon \\ 0 & \rho_i^2 \end{pmatrix}; \quad \varepsilon = 0, \text{ if } \rho_i^1 \ne \rho_i^2.$$

We denote by J the set of indices such that

$$\varphi_i^1 > \varphi_i^2, \quad i \in J.$$

If this set is empty or $J = \{l\}$ for some l, then we transform our system by the transformation $\check{Y}(\tilde{x}) = \Gamma(x)Y(\tilde{x})$, where

$$\Gamma(x) = \prod_{i=1; i \ne l}^{n} \left(\frac{x - a_l}{x - a_i} \right)^{\varphi_i^1} I. \tag{6.2.3}$$

As a result we obtain a system, satisfying the condition of the proposition. Indeed, $\Gamma(x) \sim I + o(1)$ at ∞ and

$$\tilde{Y}_i(\tilde{x}) = \Gamma Y_i(\tilde{x}) = \tilde{V}_i(x)(\tilde{x} - a_i)^{E_i}, \quad i \ne l,$$

$$\tilde{Y}_l(\tilde{x}) = \tilde{V}_l(x)(x - a_l)^{\Phi_l'}(\tilde{x} - a_l)^{E_l}, \tag{6.2.4}$$

since $\Gamma(x)$ is the scalar matrix.

If $J = \emptyset$, we can choose any index l.

Suppose that J contains $i, m, \ldots$. To prove the lemma, we only need to show that by substituting $\tilde{Y} = \Gamma Y$ one can always reduce the number of elements of J without increasing the Fuchsian weight of the system.

We pass from Y to $Y' = V_i^{-1}(a_i)Y$, which we denote again by Y. Now we have

$$V_i(a_i) = I \qquad (6.2.5)$$

in factorization (6.2.2) for Y_i. By the matrix

$$\Gamma_1(x) = \begin{pmatrix} \frac{x-a_m}{x-a_i} & 0 \\ 0 & 1 \end{pmatrix}$$

transform Y to $Y' = \Gamma_1 Y$. It follows from the form of $\Gamma_1(x)V_i(x)$ that the valuations $\tilde{\varphi}_i^j$ for Y_i' are connected with the valuations φ_i^j for Y_i by the following relations:

$$\tilde{\varphi}_i^1 = \varphi_i^1 - 1, \; \tilde{\varphi}_i^2 = \varphi_i^2,$$

$$\Delta_i' = \tilde{\varphi}_i^1 - \tilde{\varphi}_i^2 = \Delta_i - 1, \qquad (6.2.6)$$

where $\Delta_i = \varphi_1^j - \varphi_i^2$.

We consider the matrix $V_m(x)$ in factorization (6.2.2) for Y. The remaining part of the proof breaks up into two cases:

a) $v_{11}^m(a_m) \neq 0$;

b) $v_{11}^m(a_m) = 0$.

First we consider case a). In this case the factorization has the form

$$Y_m'(\tilde{x}) = \begin{pmatrix} (x-a_m)v_{11}^m & (x-a_m)v_{12}^m \\ * & v_{22}^m \end{pmatrix} (x-a_m)^{\Phi_m}(\tilde{x}-a_m)^{E_m} =$$

$$= \begin{pmatrix} v_{11}^m & (x-a_m)v_{12}^m \\ w_{21} & v_{22}^m \end{pmatrix} (x-a_m)^{\Phi_m'}(\tilde{x}-a_m)^{E_m}, \qquad (6.2.7)$$

where $\Phi_m' = \Phi_m + \begin{pmatrix} 1 & 0 \\ 0 & 0 \end{pmatrix}$, w_{21} is meromorphic at a_m

(it may have a pole of the first order there). Let $w_{21} = c/(x-a_m)+$hol.. We can again apply the procedure of Lemma 4.1.1 to obtain holomorphy of w_{21}. For this purpose it is sufficient to transform our system by the matrix

$$\Gamma_2(x) = \begin{pmatrix} 1 & 0 \\ -\frac{c}{(x-a_m)v_{11}^m(a_m)} & 1 \end{pmatrix}$$

to a Fuchsian system with the fundamental matrix $Y''(\tilde{x}) = \Gamma_2(x)Y'(\tilde{x})$ and with valuations $\check{\varphi}_i^j$. These valuations satisfy the following equalities:

$$\check{\varphi}_m^1 = \varphi_m^1 + 1, \quad \check{\varphi}_m^2 = \varphi_m^2, \quad \Delta_m'' = \Delta_m + 1. \tag{6.2.8}$$

Let us consider case b). In this case $v_{12}^m(a_m) \neq 0$ (otherwise $\det V_m(a_m) = 0$). In the similar way as in (6.2.7) we obtain

$$Y_m''(\tilde{x}) = \begin{pmatrix} (x - a_m)\tilde{v}_{11}^m & \tilde{v}_{12}^m \\ * & w_{22} \end{pmatrix} (x - a_m)^{\Phi_m''}(\tilde{x} - a_m)^{E_m}, \text{ where}$$

$$\Phi_m'' = \Phi_m + \begin{pmatrix} 0 & 0 \\ 0 & 1 \end{pmatrix}, \quad w_{22} \text{ has (in general) a pole at } a_m.$$

Let $w_{22} = c/(x - a_m) +$ hol., then by the transformation $\tilde{Y} = \Gamma_3(x)Y''$ with

$$\Gamma_3 = \begin{pmatrix} 1 & 0 \\ -\frac{c}{(x-a_m)\tilde{v}_{12}^m(a_m)} & 1 \end{pmatrix} \tag{6.2.9}$$

we get holomorphy of w_{22}. The system with the fundamental matrix $\tilde{Y}$ is still Fuchsian and its valuations $\check{\varphi}_i^j$ satisfy the relations

$$\check{\varphi}_m^1 = \varphi_m^1, \quad \check{\varphi}_m^2 = \varphi_m^2 + 1, \quad \Delta_m'' = \Delta_m - 1. \tag{6.2.10}$$

It follows from (6.2.6)–(6.2.10) that both in case a) and in case b) we have

$$\sum_{l=1}^n (\check{\varphi}_m^1 - \check{\varphi}_m^2) \leq \gamma_B, \quad \Delta_i'' = \Delta_i - 1 \tag{6.2.11}$$

for Y''.

If $\Delta_i'' > 0$ and $\Delta_m'' > 0$, we repeat the above procedure once again. It follows from (6.2.11) that after no more than $\Delta_i = \varphi_i^1 - \varphi_i^2$ steps we shall obtain a fundamental matrix $\tilde{Y} = \Gamma(x)Y$ such that (6.2.11) holds and either $\tilde{\Delta}_i = 0$ or $\tilde{\Delta}_m = 0$. This means that the number of elements of J has decreased. We find from (6.2.6)–(6.2.11) that $\tilde{Y}$ is the fundamental matrix of a Fuchsian system (6.2.1), whose Fuchsian weight does not exceed γ_B.

Consider a Fuchsian system with a fundamental matrix $Y(\tilde{x})$ such that for the corresponding Levelt's fundamental matrices of the system factorizations (6.2.4) hold true. Let $V_l(x)$ from (6.2.2) have the form

$$V_l(x) = \begin{pmatrix} 1 & c(x - a_l)^m \\ s(x - a_l)^k & 1 \end{pmatrix} (1 + o(1)). \tag{6.2.12}$$

(To obtain such the condition it is sufficient to transform $Y(\tilde{x})$ into $\tilde{Y}(\tilde{x}) = V_l^{-1}(a_l)Y(\tilde{x})$).

Lemma 6.2.1 *If in (6.2.12) $c \neq 0, m < \gamma_B$, then $\gamma_x < \gamma_B$.*

Proof. 1. If $m \leq \frac{1}{2}\gamma_B$, then

$$Y_l(\tilde{x}) = V_l'(x)(x - a_l)^{\Phi_l'}(\tilde{x} - a_l)^{E_l},$$

where $V_l'(x) = \begin{pmatrix} (x - a_l)^m & c \\ s(x - a_l)^{k+m} & (x - a_l)^{-m} \end{pmatrix}(1+o(1))$, $\Phi_l' = \Phi_l + \begin{pmatrix} -m & 0 \\ 0 & m \end{pmatrix}$.

Note that $\varphi_l^1 - m \geq \varphi_l^2 + m$ in this case, therefore the diagonal elements of Φ_l' are in nonincreasing order.

By Lemma 4.1.1 for the second column $((\tilde{v}_{12}^l, \tilde{v}_{22}^l)^t)^2$ of V_l' there exists a polynomial Q of degree m in $\frac{1}{(x-a_l)}$ such that for

$$\Gamma = \begin{pmatrix} 1 & 0 \\ Q & 1 \end{pmatrix} \quad \text{the column } \Gamma(\tilde{v}_{12}^l, \tilde{v}_{22}^l)^t$$

is holomorphic at a_l. Since $\tilde{v}_{11}^l = (x - a_l)^m(1 + o(1))$, we have that $\Gamma V_l'$ is holomorphic at a_l and

$$\det \Gamma V_l' \cdot (x - a_l)^{\Phi_l'} = \det \Gamma \det V_l' \cdot (x - a_l)^{\text{tr}\Phi_l'} =$$

$$\det V_l' \cdot (x - a_l)^{\text{tr}\Phi_l} = \det V_l \cdot (x - a_l)^{\text{tr}\Phi_l},$$

since $\det \Gamma = 1$, $\text{tr}\Phi_l' = \text{tr}\Phi_l$. Therefore, $\det V_l'(a_l) \neq 0$. Thus the system with the fundamental matrix $Y' = \Gamma Y$ is Fuchsian. It has form (6.2.4) and its Fuchsian weight $\gamma_{B'}$ satisfies the following relations:

$$\gamma_{B'} = \gamma_B - 2m < \gamma_B.$$

Thus

$$\gamma_x \leq \gamma_{B'} < \gamma_B.$$

2. To investigate the case $\frac{1}{2}\gamma_B < m < \gamma_B$ we need the following lemma.

Lemma 6.2.2 *Let a Fuchsian system (6.2.1) be reduced to (6.2.4), (6.2.12) and let $\gamma_B > 0$. If all monodromy matrices G_i are non-diagonalizable, then there is an $i \neq l$ such that the value at a_i of the element $v_{11}^i(x)$ of the matrix $V_i(x)$ in (6.2.4) is non-zero:*

$$v_{11}^i(a_i) \neq 0. \tag{6.2.13}$$

---

^2t means transposition here

Proof. By assumption, it follows that all matrices E_i in (6.2.4) have the form of Jordan blocks. Recall here that by (2.2.31)

$$B^j = V_j(a_j)L_j(a_j)V_j^{-1}(a_j), \qquad (6.2.14)$$

where $L_j(x) = \Phi_j(x) + (x - a_j)^{\Phi_j} E_j(x - a_j)^{-\Phi_j}$.

Suppose that $v_{11}^i(a_i) = 0$ for $i = 1, \ldots, n$, $i \neq l$. Then we find from (6.2.14) and (6.2.4) that

$$B^i = V_i(a_i)E_iV_i^{-1}(a_i) = \begin{pmatrix} \rho_i & 0 \\ * & \rho_i \end{pmatrix}.$$

Since $\gamma_B > 0$, it follows from (6.2.14) that $B^l = \operatorname{diag}(\beta_l^1, \beta_l^2)$. Indeed,

$$L_l(a_l) = \lim_{x \to a_l} \begin{pmatrix} \beta_l^1 & (x - a_l)^{\varphi_l^1 - \varphi_l^2} \\ 0 & \beta_l^2 \end{pmatrix} = \begin{pmatrix} \beta_l^1 & 0 \\ 0 & \beta_l^2 \end{pmatrix}.$$

Consequently, we find from Theorem 2.2.2 (the sum of exponents equals zero)) that

$$\sum_{i=1, i \neq l}^n \rho_i + \beta_l^1 = \sum_{i=1, i \neq l}^n \rho_i + \beta_l^2 = 0$$

and so $\beta_l^1 = \beta_l^2$. Since $\rho_l^1 = \rho_l^2$, we have $\varphi_l^1 = \varphi_l^2$, which contradicts the assumption that $\gamma_B > 0$.

Let us return to the second part of Lemma 6.2.2. Let now $\frac{1}{2}\gamma_B < m < \gamma_B$. There are two possibilities for E_l in (6.2.4):

a) E_l is a diagonal matrix,

b) E_l is a Jordan block.

In case a) repeat the procedure of the item 1 for $Y_l(\tilde{x})$, and at the end permute the columns of Y'. As a result we obtain a system with factorization (6.2.4), where E_l' is a diagonal (as it was before the permutation) and $\Phi_l' = \operatorname{diag}(\tilde{\varphi}_l^1, \tilde{\varphi}_l^2)$, where $\tilde{\varphi}_l^1 = \varphi_l^1 + m$, $\tilde{\varphi}_l^2 = \varphi_2^1 - m$. Since

$$\tilde{\varphi}_l^1 - \tilde{\varphi}_l^2 = \varphi_l^1 - \varphi_l^2 + 2m = 2m - \gamma_B > 0,$$

we still have a Levelt's factorization in $\tilde{U}_l^*$, therefore the system with the fundamental matrix Y' is Fuchsian at a_l and

$$\gamma_{B'} = 2m - \gamma_B < 2\gamma_B - \gamma_B < \gamma_B.$$

Thus, $\gamma_\chi < \gamma_B$.

Let us consider case b). If E_i is a diagonal matrix for some $i \neq l$, then exchanging the columns of $Y_i(\tilde{x})$ if necessary we find that (6.2.13) holds true for $V_i(x)$ from (6.2.4). By virtue of Lemma 6.2.2, if E_j is a Jordan block for any $j \neq l$, then there is also an $i \neq l$ such that (6.2.13) holds.

Transform the fundamental matrix $Y(\tilde{x})$ of our system to $Y'(\tilde{x}) = \Gamma_1(x)Y(\tilde{x})$ by

$$\Gamma_1(x) = \begin{pmatrix} \left(\frac{x-a_i}{x-a_l}\right)^{2m-\gamma_B} & 0 \\ 0 & 1 \end{pmatrix}.$$

Under this transformation the matrix $Y_l'(\tilde{x}) = \Gamma_1(x)Y_l(\tilde{x})$ has the form

$$Y_l'(\tilde{x}) = V_l'(x)(x - a_l)^{\Phi_l - C}(\tilde{x} - a_l)^{E_l},$$

where $C = \operatorname{diag}(2m - \gamma_B, 0)$, $V_l'(x)$ is holomorphically invertible at a_i, and its first row has the form

$$((a_l - a_i)^{2m-\gamma_B}, \; c'(x - a_l)^{\gamma_B - m})(1 + o(1)), \; c' \neq 0.$$

The valuation $\tilde{\varphi}_l^j$ for Y_l' satisfy the following equalities: $\tilde{\varphi}_l^1 = \varphi_l^1 - (2m - \gamma_B)$ and $\tilde{\varphi}_l^2 = \varphi_l^2$. Let us note that $\tilde{\varphi}_l^1 - \tilde{\varphi}_l^2 = 2(\gamma_B - m)$. Thus, applying the procedure of the item 1 to Y', we obtain $Y'' = \Gamma_2 Y'$ and the corresponding system with valuations $\breve{\varphi}_i^j$ such that

$$\breve{\varphi}_l^1 = \breve{\varphi}_l^2. \tag{6.2.15}$$

This procedure does not change the form

$$(\tilde{v}_{11}^i(x - a_i)^{2m-\gamma_B}, \; \tilde{v}_{12}^i(x - a_i)^{2m-\gamma_B}), \; \tilde{v}_{11}^i(a_i) \neq 0$$

of the first row of the matrix $\Gamma_2 V_i'$. Therefore,

$$Y_i''(\tilde{x}) = \Gamma_2(x)\Gamma_1(x)Y_i(\tilde{x}) = V_i''(x)(x - a_i)^D(\tilde{x} - a_i)^{E_i},$$

where $D = (2m - \gamma_B, 0)$, $V_i''(x) = \begin{pmatrix} \tilde{v}_{11}^i(x) & \tilde{v}_{12}^i(x)(x - a_i)^{2m-\gamma_B} \\ \tilde{w}_{21}^i(x) & \tilde{v}_{22}^i(x) \end{pmatrix}$,

$\tilde{v}_{11}^i(a_i) \neq 0$, $\tilde{v}_{12}^i, \tilde{v}_{22}^i$ are holomorphic at a_i, $\tilde{v}_{22}^i(a_i) \neq 0$, $\tilde{w}_{21}^i$ in general has a pole at most of the order $(2m - \gamma_B)$.

Applying to the first column of V_i'' the procedure of item 1, we obtain a Fuchsian system with a fundamental matrix $Y^0(\tilde{x}) = \Gamma_3 Y''(\tilde{x})$ and with valuations φ_i^j, which satisfy the following conditions:

$$\varphi_t^1 = \varphi_t^2 = 0, \; t = 1, \ldots, n, \; t \neq l, \; t \neq i,$$

$$\varphi_i^1 - \varphi_i^2 = 2m - \gamma_B.$$

Since by (6.2.15) $\varphi_l^1 = \varphi_l^2$, we obtain $\gamma_B' = 2m - \gamma_B < \gamma_B$ for this system, therefore

$$\gamma_\chi < \gamma_B.$$

Return to the proof of Theorem 6.2.1. By Proposition 6.2.1 there exists a Fuchsian system (6.2.1) such that for its valuations the following equalities hold:

$$\varphi_i^1 = \varphi_i^2 = 0, \ i = 1, \ldots, n, \ i \neq l,$$

$$\varphi_l^1 - \varphi_l^2 = \gamma_B = \gamma_\chi.$$

By a constant matrix S transform this system to a system such that the matrix $V_l(x)$ from factorization (6.2.4) for a fundamental matrix of this system has form (6.2.12). Then by Lemma 6.2.2 either $c = 0$ or $m \geq \varphi_l^1 - \varphi_l^2$. But in both these cases we have

$$Y_l(\tilde{x}) = V_l(x - a_l)^{\Phi_l}(\tilde{x} - a_l)^{E_l} = (x - a_l)^{\Phi_l}\tilde{V}_l(\tilde{x} - a_l)^{E_l}, \qquad (6.2.16)$$

where $\tilde{V}_l$ is still holomorphically invertible at a_l. Indeed, for elements $\tilde{v}_{mt}^i$ of this matrix we have

$$\tilde{v}_{11}^i = v_{11}^i, \ \tilde{v}_{21}^i = v_{21}^i(x - a_l)^{\gamma_\chi},$$

$$\tilde{v}_{22}^i = v_{22}^i, \ \tilde{v}_{12}^i = v_{12}^i(x - a_l)^{-\gamma_\chi} = c(x - a_l)^{m-\gamma_\chi}(1 + o(1))$$

and under the assumption ($c = 0$ or $m \geq \gamma_\chi$) $\tilde{v}_{12}^i$ is still holomorphic at a_l.

Due to Section 5.1 $Y(\tilde{x})$ determines a meromorphic section of the canonical extension R^0 for the representation χ such that the section is holomorphically invertible off a_l. And (6.2.16) means that the collection of diagonal elements of Φ_l presents the splitting type of G^0. Thus,

$$\gamma(0) = \gamma_\chi.$$

Theorem 6.2.2 *Any representation χ of dimension $p = 3$ with any points a_1, a_2, a_3 can be realized as the monodromy representation of some Fuchsian system.*

Proof. Due to Theorem 6.1.1 we can reduce our observation to a reducible representation χ of dimension $p = 3$ such that each of the matrices G_i can be reduced to a Jordan block and the corresponding two-dimensional subrepresentation or quotient representation χ_2 is irreducible. Since χ is reducible, it follows that for the corresponding one-dimensional subrepresentation or quotient representation the number

$$\sum_{i=1}^{3} \rho_i$$

is integer. Indeed, let $G_1, \ldots, G_n$ have the form

$$G_i \begin{pmatrix} \lambda_i & & \\ 0 & & * \\ 0 & & \end{pmatrix} \text{ or } G_i = \begin{pmatrix} & & * \\ 0 & 0 & \lambda_i \end{pmatrix}.$$

In both these cases we get $\lambda_1 \cdots \lambda_n = 1$ and therefore $r = \rho_1 + \cdots + \rho_n$ is integer.

Consider a Fuchsian system with the monodromy χ_2. By Theorem 2.2.2, applied to this system we obtain

$$\sum_{i=1}^{n} (\rho_i + \tilde{\varphi}_i^1) + \sum_{i=1}^{n} (\rho_i + \tilde{\varphi}_i^2) = 0, \quad \text{therefore} \quad \sum_{i=1}^{n} (\tilde{\varphi}_i^1 + \tilde{\varphi}_i^2) = -2r,$$

$$\text{and } \gamma_{B'} = \sum_{i=1}^{n} (\tilde{\varphi}_i^1 - \tilde{\varphi}_i^2) = -2r - 2 \sum_{i=1}^{n} \tilde{\varphi}_i^2$$

for Fuchsian weight $\gamma_{B'}$ of the system. The latter equality means that this weight is an even number. Thus, γ_{χ_2} is even too. Hence, by Proposition 4.2.1 and Theorem 6.2.1 we have

$$\gamma_{\chi_2} = \gamma(0) \le 1,$$

therefore $\gamma_{\chi_2} = 0$. Thus, by Theorem 6.2.1 the representation χ can be realized as the monodromy representation of some Fuchsian system.

6.3   Properties of Fuchsian weight

In this section we continue the investigation of Fuchsian weight.

Proposition 6.3.1 *For any representation χ of dimension $p = 2$, the inequalities*

$$\gamma_B \ge 0, \quad \gamma_\chi \ge 0 \tag{6.3.1}$$

hold, and the parity of γ_B and γ_χ coincides with that of $\sum_{i=1}^{n} tr E_i$. If χ is a commutative representation that cannot be decomposed into a direct sum of one-dimensional representations, then $\gamma_\chi = 0$.

Proof. The first part of the proposition follows from the definitions of γ_B and γ_χ and the fact that by Theorem 2.2.2

$$0 = \sum_{i=1}^{n}(\beta_i^1 + \beta_i^2) = \sum_{i=1}^{n} \mathrm{tr} E_i + \sum_{i=1}^{n} \varphi_i^1 + \sum_{i=1}^{n} \varphi_i^2,$$

$$\sum_{i=1}^{n} \varphi_i^1 + \sum_{i=1}^{n} \varphi_i^2 = -\sum_{i=1}^{n} \mathrm{tr} E_i, \text{ and so}$$

$$\gamma_B = \sum_{i=1}^{n}(\varphi_i^1 - \varphi_i^2) = -\sum_{i=1}^{n} \mathrm{tr} E_i - 2\sum_{i=1}^{n} \varphi_i^2.$$

If χ is a commutative representation that cannot be decomposed into a direct sum of one-dimensional representations, then the matrices G_j can be simultaneously reduced to the form

$$\begin{pmatrix} \lambda_i & * \\ 0 & \lambda_i \end{pmatrix}.$$

Indeed, since χ is commutative, we have that χ is reducible. Thus, all $\chi(\sigma_i)$ can be reduced to the form

$$G_i = \begin{pmatrix} \lambda_i^1 & \mu_i \\ 0 & \lambda_i^2 \end{pmatrix},$$

where $\mu_i \neq 0$, $\lambda_i^1 = \lambda_i^2$ for some i. (Otherwise one of them can be reduced to a diagonal matrix with different eigenvalues. Since G_i commute it follows that all of them are such matrices and therefore χ is decomposable). Thus, from commutativity of χ we get $\lambda_j^1 = \lambda_j^2$ for all j). Therefore,

$$Y(\tilde{x}) = (x - a_1)^{-\frac{1}{2}\sum_{i=1}^{n} \mathrm{tr} E_i} \prod_{i=1}^{n}(\tilde{x} - a_i)^{E_i} \tag{6.3.2}$$

is the fundamental matrix for a Fuchsian system (6.2.1) with the given monodromy and with weight 0. Thus $\gamma_\chi = 0$.

To calculate Fuchsian weight of a representation we need first of all the procedure for calculating the number m in (6.2.12) for factorization (6.2.4). For the concrete examples this procedure was presented in Chapters 2,3. Here we present slightly different way.

Lemma 6.3.1 *Let (1.2.1) be a Fuchsian system reduced to form (6.2.4), (6.2.12). Then the matrix A of the system can be written as*

$$\left(\begin{array}{cc} \frac{\beta_1^1}{x - a_l} & c(m - \beta_1^1 + \beta_1^2)(x - a_l)^{m-1} + \varepsilon(x - a_l)^{\gamma_B - 1} \\ s(k + \beta_1^1 - \beta_1^2)(x - a_l)^{k-1} & \frac{\beta_1^2}{x - a_l} \end{array} \right).$$

(6.3.3)

$$(1 + o(1))$$

in a neighborhood of a_l, where ε is defined in (6.2.2).

Proof. We find from (6.2.12) that

$$V_l^{-1} = \left(\begin{array}{cc} 1 & -c(x - a_l)^m \\ -s(x - a_l)^k & 1 \end{array} \right) (1 + o(1)),$$

$$\frac{dV_l}{dx} = \left(\begin{array}{cc} \alpha(x - a_l)^{t_1} & cm(x - a_l)^{m-1} \\ sk(x - a_l)^{k-1} & \delta(x - a_l)^{t_2} \end{array} \right) (1 + o(1)),$$

where $t_1 \geq 0$ and $t_2 \geq 0$. Formula (6.3.3) follows now from (2.2.29). Recall that in our case this formula has the form

$$A = \frac{dV_l}{dx} V_l^{-1} + \frac{V_l}{x - a_l} (\Phi_l + (x - a_l)^{\Phi_l} E_l (x - a_l)^{-\Phi_l}) V_l^{-1}$$

(6.3.4)

and it yields (6.2.14).

Lemma 6.3.2 *Suppose that an element a_{pq} of the matrix A of a Fuchsian system (1.2.1) has a decomposition of the form*

$$a_{pq} = \frac{1}{x - a_l} c_{-1}^{pq} + c_0^{pq} + \cdots$$

(6.3.5)

in a neighborhood of a_l. Then the elements b_{pq}^i of the matrices B^i in (6.2.1) are connected with the numbers c_k^{pq} by the following relations:

$$b_{pq}^l = c_{-1}^{pq},$$

$$\sum_{i=1, i \neq l}^{n} b_{pq}^i \frac{1}{(a_i - a_l)^r} = -c_{r-1}^{pq}, \quad r = 1, \ldots$$

(6.3.6)

Proof. From (1.2.1) and (6.2.1) we have

$$A - B^l \frac{1}{x - a_l} = \sum_{i=1, i \neq l}^{n} B^i \frac{1}{x - a_i}.$$

Expanding the right hand side of this formula into series in $\frac{1}{x - a_l}$, we get (6.3.5).

Proposition 6.3.2 *Fuchsian weight γ_χ of any representation χ satisfies the inequality $\gamma_\chi < n - 1$. If χ is an irreducible representation, then*

$$\gamma_\chi \leq n - 2.$$

Proof. The last inequality follows from Theorem 6.2.1 and Proposition 4.2.1. Indeed,

$$\gamma_\chi = \gamma(0) \leq \frac{1}{2}(n - 2)p(p - 1) = n - 2.$$

Let χ be reducible. Assume that $\gamma_\chi > n - 2$. Then, according to Lemma 6.2.1, $m \geq n - 1$, where m is the number from (6.2.12). In this case it follows from Lemma 6.3.1 that the element a_{12} of the matrix A of the system can be written as $a_{12} = o((x - a_l)^{n-3})$ in a neighborhood of a_l. It also follows from (6.2.14), (1.2.3), and the condition $\gamma_\chi > 0$ that $b_{12}^l = 0$ and

$$b_{12}^1 + \cdots + b_{12}^n = 0.$$

Therefore, we find from Lemma 6.3.2 that the numbers b_{12}^i for $i \neq l$ satisfy the system of $n - 1$ equations (6.3.6)

$$\sum_{i=1, i \neq l}^{n} b_{12}^i \frac{1}{(a_i - a_l)^r} = 0, \quad r = 0, \ldots, n - 2,$$

whose determinant is the Vandermonde determinant of the numbers

$$\frac{1}{a_1 - a_l}, \ldots, \frac{1}{a_{l-1} - a_l}, \frac{1}{a_{l+1} - a_l}, \ldots, \frac{1}{a_n - a_l},$$

and so it is non-zero. Thus, $b_{12}^i = 0$ for $i = 1, \ldots, n$. Consequently, all the matrices B^i of (6.2.1) have the lower-triangular form, and we find that

$$b_{11}^l = \rho_l^1 + \varphi_l^1, \; b_{22}^l = \rho_l^2 + \varphi_l^2, \; b_{11}^i = \rho_i^1, \; b_{22}^i = \rho_i^2, \; i \neq l. \tag{6.3.7}$$

Since (1.2.3) implies that $\sum_{i=1}^{n} \rho_i^j + \varphi_i^j = 0$ for $j = 1, 2$, and (2.2.2) implies that $0 \leq \operatorname{Re} \sum_{i=1}^{n} \rho_i^j \leq n - 1$ for $j = 1, 2$, it follows from the last two equalities that $\varphi_l^1 - \varphi_l^2 \leq n - 1$, which in conjunction with the assumption that $\gamma_\chi > n - 2$ yields the equality $\gamma_\chi = n - 1$.

Since the matrices B^i have the lower-triangular form, it follows that the representation χ is reducible in the case in question. So, we get another proof of inequality $\gamma_\chi \leq n - 2$ for irreducible representations.

Proposition 6.3.3 *For any points $a_1, \ldots, a_n$ and any number γ that satisfies the inequalities*

$$0 < \gamma \le n - 2, \tag{6.3.8}$$

there is an irreducible representation χ such that $\gamma_\chi = \gamma$ and the monodromy matrices $\chi(\sigma_i)$ are non-diagonalizable.

Proof. Suppose, that $a_1 = 0$ and ∞ is not among the points in D. (We can always make sure that this is the case by applying a linear fractional transformation of the Riemann sphere.)

1. First, we shall prove the proposition in the case of an even $\gamma = 2\gamma'$. Let us consider the following two systems of equations for unknown $d_2, \ldots, d_n$ and $c_2, \ldots, c_n$:

$$\sum_{i=2}^{n} d_i \frac{1}{a_i^r} = \delta_{r,\gamma}, \tag{6.3.9}$$

and

$$\sum_{i=2}^{n} c_i \frac{1}{a_i^r} = x^2 \delta_{r,n-2}, \tag{6.3.10}$$

$$\text{where } r = 0, 1 \ldots, n \text{ and } \delta_{i,j} = \begin{cases} 0, & i \ne j, \\ 1, & i = j. \end{cases}$$

Since the determinants of (6.3.9) and (6.3.10) do not vanish, there is a j such that $d_j \ne 0$, d being the solution of (6.3.9). But any solution c of (6.3.10) has the form $c = z^2 t$ with $t_i \ne 0$ for $i = 2, \ldots, n$. Thus, we can choose the values of the roots $\sqrt{d_i t_i}$ so that $s = \sum_{i=2}^{n} \sqrt{d_i t_i} \ne 0$. Let us set $z = -\gamma/(2s) = -\gamma'/s$. Then

$$\sum_{i=2}^{n} \sqrt{d_i c_i} = zs = -\frac{\gamma}{2} = -\gamma',$$

and so the matrices

$$B^1 = \begin{pmatrix} \gamma' & 0 \\ 0 & -\gamma' \end{pmatrix}, \; B^i = \begin{pmatrix} \sqrt{d_i c_i} & d_i \\ -c_i & -\sqrt{d_i c_i} \end{pmatrix} \tag{6.3.11}$$

satisfy (1.2.3).

Let us consider a Fuchsian system (6.2.1) with matrices (6.3.11). From (6.2.14) and the fact that diagonal elements of $L_l(a_i)$ are β_l^j we get

$$\beta_i^j = \rho_i^j = \varphi_i^j = 0, \; i \ne 1, \; j = 1, 2, \tag{6.3.12}$$

$$\beta_1^1 = \varphi_1^1 = \gamma', \ \beta_1^2 = \varphi_1^2 = -\gamma'$$

for the exponents of the system. Indeed, the matrices B^i, $i \neq 1$ have only zero eigenvalues, therefore we get the first equality in (6.3.12). The second equality in (6.3.12) follows from the form of the matrix B^1 in (6.3.11).

Let us consider factorization (6.2.2) at $a_1 = 0$:

$$Y_1(\tilde{x}) = V_1(x)x^{\Phi_1}\tilde{x}^{E_1}, \ \tilde{x} \in \tilde{U}_1^*. \tag{6.3.13}$$

It follows from (6.3.12) that

$$\Phi_1 = \begin{pmatrix} \gamma' & 0 \\ 0 & -\gamma' \end{pmatrix}, \ E_1 = \begin{pmatrix} 0 & \alpha \\ 0 & 0 \end{pmatrix}. \tag{6.3.14}$$

From (6.3.11) and (6.2.14) we find that

$$\begin{pmatrix} \gamma' & 0 \\ 0 & -\gamma' \end{pmatrix} = V_1(0) \begin{pmatrix} \gamma' & 0 \\ 0 & -\gamma' \end{pmatrix} V_1^{-1}(0),$$

and so

$$V_1(0) = \begin{pmatrix} u_{11} & 0 \\ 0 & u_{22} \end{pmatrix}.$$

Consider the matrix

$$Y_1' = Y_1 \begin{pmatrix} u_{11}^{-1} & 0 \\ 0 & u_{22}^{-1} \end{pmatrix}.$$

Factorization (6.3.13) holds true for the matrix with the same Φ_1, with

$$E_1 = \begin{pmatrix} 0 & \varepsilon \\ 0 & 0 \end{pmatrix}, \ \varepsilon = \alpha\frac{u_{11}}{u_{22}}$$

and with $V_1(x)$ of form (6.2.11). Denote again Y_1' by Y_1. From (6.3.9) and Lemma 6.3.2 for the matrix A of the constructed system (1.2.1), (6.2.1), (6.3.11) we get

$$a_{12} = x^{\gamma-1} + o(x^{\gamma-1}). \tag{6.3.15}$$

It follows from (6.3.15) and Lemma 6.3.1 that the number m for $V_1(x)$ in (6.3.11) satisfies the inequality $m \geq \gamma$. From this inequality, Theorem 6.2.1 and Lemma 6.2.2 we find that $\gamma_B = \gamma_\chi = \gamma$ for the constructed system.

It remains to prove that the monodromy matrices $G_1, \ldots, G_n$ of the constructed system are non-diagonalizable and the monodromy representation of the system is irreducible.

By virtue of the definition of E_i to prove that $G_1, \ldots, G_n$ are non-diagonalizable it is sufficient to show that so are $E_1, \ldots, E_n$. It follows from (6.2.14) and (6.3.12) that for $i \neq 1$ the matrices L_i in (6.2.14) coincide with E_i. Thus, by virtue of (6.2.14) the matrices E_i are non-diagonalizable since the matrices B^i in (6.3.11) are non-diagonalizable for $i \neq 1$.

We claim that E_1 is also non-diagonalizable. We observe that in (6.3.3) if $c \neq 0$, then $m = \gamma$, and the expression $c(m - \beta_1^1 + \beta_1^2)$ vanishes by virtue of (6.3.12) and so does this expression for $c = 0$. Therefore, if $m \geq \gamma$, then the element a_{12} in (6.3.3) is of the form

$$a_{12} = \varepsilon x^{\gamma_B - 1} + o(x^{\gamma_B - 1}).$$

Comparing the last equality with (6.3.15) and using the equality $\gamma_B = \gamma_X = \gamma$, which was proved earlier, we find that $\varepsilon = 1$ and $\alpha = u_{22}/u_{11} \neq 0$, which means that E_1 is a non-diagonalizable matrix.

However, if the monodromy representation of the constructed system (6.2.1), (6.3.11) were reducible, then the fact that $G_1, \ldots, G_n$ are non-diagonalizable would imply that the representation is commutative. In this case we would find from Proposition 6.3.1 that $\gamma_X = 0$, contrary to the equality $\gamma_X = \gamma > 0$ already proved.

2. We consider the case when $\gamma = 2\gamma'' + 1$ is odd. In this case the proof can be carried out exactly as in the case of even γ but with the matrices B_i in (6.3.11), replaced by

$$B_1 = \begin{pmatrix} -n + 2 + \gamma'' & 0 \\ 0 & -n + 1 - \gamma'' \end{pmatrix},$$

$$B_i = \begin{pmatrix} \sqrt{d_i c_i} + \frac{2n-3}{2n-2} & d_i \\ -c_i & -\sqrt{d_i c_i} + \frac{2n-3}{2n-2} \end{pmatrix}, \quad i \neq 1,$$

with Φ_1 replaced by

$$\Phi_1 = \begin{pmatrix} -n + 2 + \gamma'' & 0 \\ 0 & -n + 1 - \gamma'' \end{pmatrix},$$

and with the exponents in (6.3.12) replaced by

$$\beta_i^j = \rho_i^j = \frac{2n - 3}{2n - 2}, \quad i \neq 1, \quad j = 1, 2,$$

$$\beta_1^1 = \varphi_1^1 = -n + 2 + \gamma'', \quad \beta_1^2 = \varphi_1^2 = -n + 1 - \gamma''.$$

Remark 6.3.1 *Since Y_1 in (6.3.13) has the form (6.2.4), (6.2.12), it follows that*

$$\varphi_1(y_{11}^1) = \gamma', \quad \varphi_1(y_{12}^1) \geq 0, \quad \varphi_1(y_{21}^1) \geq \gamma', \quad \varphi_1(y_{22}^1) = -\gamma'$$

for the elements y_{ij}^1 of Y_1.

The system (2.1.4) of Chapter 2 is the example of a system, satisfying the conditions of Proposition 6.3.3 for $n = 4$, $p = 3$, $\gamma = 2$.

6.4   Instability of Fuchsian weight

In this section we use the following notation: S_a for the space $\bar{\mathbb{C}} \setminus \{a_1, \ldots, a_n\}$, $\tilde{S}_a$ for a universal covering of S_a, and Δ_a for the corresponding group of deck transformations. We also equip by index a the generators $\sigma_i^a \in \Delta_a$ and the ways $\beta_1^a, \ldots, \beta_n^a$, which take part in identification (2.2.42).

We denote by F_a a group with n generators $h_1, \ldots, h_n$ satisfying the identity relation $h_1 \cdot \ldots \cdot h_n = e$ and we denote by κ_a the isomorphism

$$\kappa_a : \Delta_a \longrightarrow F_n \tag{6.4.1}$$

such that $\kappa_a(\sigma_i^a) = h_i$.

Any representation χ can be written in the form $\chi = \chi' \circ \kappa_a$, where

$$\chi' : F_n \longrightarrow GL(p, C) \tag{6.4.2}$$

is a representation of the group F_n. In what follows we shall denote χ by $\chi(a)$, where $a = (a_1, \ldots, a_n)$.

The proof of the following statement is straightforward.

Lemma 6.4.1 *If for two sequences of points* $a = (a_1, \ldots, a_n)$ *and* $b = (b_1, \ldots, b_n)$ *there is a linear fractional transformation* $r : \bar{\mathbb{C}} \longrightarrow \bar{\mathbb{C}}$ *such that*

$$r(a_i) = b_i, \quad r(\beta_i^a) = \beta_i^b, \quad i = 1, \ldots, n,$$

then $\gamma_{\chi(a)} = \gamma_{\chi(b)}$.

Theorem 6.4.1 *Let the Fuchsian weight* $\gamma_{\chi(a)}$ *of a representation* $\chi(a)$ *with non-diagonalizable monodromy matrices* $\chi'(h_1), \ldots, \chi'(h_n)$ *be greater than one. Then there are an* $\varepsilon > 0$ *and an index* l *such that the inequality*

$$\gamma_{\chi(a')} < \gamma_{\chi(a)} - 1$$

holds for any sequence of points $a' = (a_1, \ldots, a_{l-1}, a_l + t, a_{l+1}, \ldots, a_n)$ *with* $0 < |t| < \varepsilon$.

Proof. Let us consider a Fuchsian system (6.2.1), reduced to form (6.2.4), (6.2.12) with the monodromy χ and with Fuchsian weight $\gamma = \gamma_{\chi(a)}$. It follows from Lemma 6.2.2 that in (6.2.12) either $c \neq 0$ and $m \geq \gamma$ or $c = 0$. We use a linear fractional map to transform a_l into 0 and ∞ into ∞. We denote the resulting sequence again by $a_1, \ldots, a_n$.

We consider the isomonodromic deformation

$$\frac{dy}{dx} = \left[\left(\frac{B^1(t)}{x - t} + \sum_{i=2}^{n} \frac{B^1(t)}{x - a_i} \right) \right] y \qquad (6.4.3)$$

of the constructed system, where

$$\begin{cases} \frac{dB^i(t)}{dt} = \frac{1}{a_i - t}[B^i(t), B^1(t)], \\ B^i(0) = B^i, \quad \sum_{i=1}^{n} B^i(t) = 0, \end{cases} \qquad (6.4.4)$$

and t varies over a small neighborhood of 0. For sufficiently small t (6.4.3) has the same monodromy as the original system (6.2.1) (see [Sch]).

Let the matrix Φ_l in (6.2.4) has the form $\Phi_l = \mathrm{diag}(b + \gamma_1, b - \gamma_2)$, where $\gamma_1 + \gamma_2 = \gamma$, $\gamma_1 = \left[\frac{\gamma+1}{2} \right]$ ([x] means "integer part of x").

From (6.2.12) and (6.2.14) we get

$$B^1(0) = \begin{pmatrix} b + \gamma_1 & 0 \\ 0 & b - \gamma_2 \end{pmatrix}.$$

Hence, it follows from (6.4.4) that

$$\frac{dB^i(0)}{dt} = \frac{1}{a_i} \begin{pmatrix} 0 & -\gamma b_{12}^i \\ \gamma b_{21}^i & 0 \end{pmatrix}$$

and

$$B^i(t) = \begin{pmatrix} b_{11}^i + o(t) & b_{12}^i - \frac{\gamma}{a_i} b_{12}^i t + o(t) \\ b_{21}^i + \frac{\gamma}{a_i} b_{21}^i t + o(t) & b_{22}^i + o(t) \end{pmatrix}. \qquad (6.4.5)$$

From this and again from (6.4.4) we have

$$B^1(t) = -\sum_{i=2}^{n} B^i(t) = \begin{pmatrix} b + \gamma_1 + o(t) & o(t) \\ o(1) & b - \gamma_2 + o(t) \end{pmatrix} \qquad (6.4.6)$$

since, by Lemmas 6.3.1, 6.3.2 and Proposition 6.3.2, the equality

$$\sum_{i=2}^{n} \frac{b_{12}^i}{a_i^r} = h\delta_{r,\gamma}, h \neq 0, r = 0, \ldots, \gamma, 2 \leq \gamma \leq n - 2 \qquad (6.4.7)$$

holds.

Factorization (6.2.4) for system (6.4.3) in $\tilde{U}_t^*$ is of the form

$$Y_1(\tilde{x}, t) = V_1(x, t)(x - t)^{\Phi_1}(\tilde{x} - t)^{E_1}.$$

It follows from (6.2.14) that $V_1(t, t)B^1(0) = B^1(t)V_1(t, t)$. Thus, as in the proof of Proposition 6.3.3 we may assume that

$$V_1(t, t) = \begin{pmatrix} 1 + x_1 t + o(t) & o(t) \\ o(1) & 1 + x_2 t + o(t) \end{pmatrix}.$$

We shall reduce the system (6.4.3) to the form (6.2.4), (6.2.12). With this end in view we pass from its fundamental matrix Y to $Y' = V^{-1}(t, t)Y$. The matrices of coefficients of the new system have the form

$$\begin{cases} \tilde{B}^1(t) = \begin{pmatrix} b + \gamma_1 & 0 \\ 0 & b - \gamma_2 \end{pmatrix}, \\ \tilde{B}^i(t) = V_1^{-1}(t, t)B^i(t)V_1(t, t) = \\ \qquad = \begin{pmatrix} * & b_{12}^i - (x_1 - x_2)b_{12}^i t - \gamma \frac{b_{12}^i}{a_i} t + o(t) \\ * & * \end{pmatrix}. \end{cases} \tag{6.4.8}$$

From (6.4.8) and (6.4.5) we get

$$\sum_{i=2}^n \frac{\tilde{b}_{12}^i}{a_i^r} = o(t),\ 0 \le r \le \gamma - 2,$$

$$\sum_{i=2}^n \frac{\tilde{b}_{12}^i}{a_i^{\gamma-1}} = -\gamma h t + o(t)$$

for elements $\tilde{b}_{12}^i$ of the matrices $\tilde{B}^i$.

Thus, according to Lemma 6.3.1, we find that for any sufficiently small t, $c_{\gamma-2}^{12} \ne 0$ in formula (6.3.5) for the matrix $\tilde{A}$ of the new system in a neighborhood of $x = t$. Hence, using (6.3.3) and the equality $\gamma_B = \gamma$, we find that $c \ne 0$ and $m \le \gamma - 1$ in (6.3.3), (6.2.12) for the new system. From Lemma 6.2.2 we deduce that $\gamma_{\chi(a')} < \gamma_{\chi(a)}$. Proposition 6.3.1 implies that the parity of $\gamma_{\chi(a)}$ is the same as that of $\gamma_{\chi(a')}$. Therefore, $\gamma_{\chi(a')} < \gamma_{\chi(a)} - 1$, where $a' = (a_1 + t, a_2, \ldots, a_n)$ and t is sufficiently small. To complete the proof of the theorem, we only need to apply the transformation inverse to the transformation introduced at the beginning of the proof and use Lemma 6.4.1.

Remark 6.4.1 *Due to Theorem 6.2.1 the instability of the Fuchsian weight of a representation is equivalent to the unstability of a vector bundle on the Riemann sphere with splitting type $(c_1, c_2), c_1 - c_2 \ge 2$. It is well known (see [Boj]), that under "small" analytic transformation each such a bundle is transformed to a bundle with splitting type $(c_1', c_2'), c_1' - c_2' \le 1$.*

6.5   The theorem of realization

Theorem 6.1.1 gives the complete answer to Hilbert's 21st problem in dimension $p = 3$. The counterexample of Chapter 2 shows that this problem really has a negative solution (in general) for $p = 3$. The following statement provides an existence of a negative solution for all given $a_1, \ldots, a_n$, $n > 3$, $p \geq 3$.

Theorem 6.5.1 *For any $n > 3$, any sequence of points $a_1, \ldots, a_n$, and any $p \geq 3$, there is a representation χ for which there are no Fuchsian systems that realize the representation.*

Proof. **1.** First we shall prove the theorem for $p = 3$, for which it is sufficient to construct a representation χ, that satisfies the conditions of Theorem 6.1.1.

Let us consider the system (6.2.1), (6.3.11). Since the number n of points is greater than three, there are row vectors $b_2, \ldots, b_n$ of two components such that $b_j \neq 0$ for $j = 2, \ldots, n$, and

$$\sum_{j=2}^{n} b_j = 0, \quad \operatorname{rank} \tilde{B}^j = 2, \ j = 2, \ldots, n, \tag{6.5.1}$$

where we set

$$\tilde{B}^1 = \begin{pmatrix} 0 & x^{-\gamma'} & 0 \\ 0 & & \\ 0 & & B^1 \end{pmatrix}, \quad \tilde{B}^i = \begin{pmatrix} 0 & b_i \\ 0 & \\ 0 & B^i \end{pmatrix}, \ i > 1. \tag{6.5.2}$$

Here B^i are the matrices given by (6.3.11), and $\gamma' = \gamma_\chi/2$, where γ_χ is the Fuchsian weight of the monodromy representation χ of (6.2.1), (6.3.11) and $\gamma_\chi > 0$.

Let us consider the system (6.2.1), (6.5.2). By virtue of (6.5.1), the system has no a singularity at ∞. We claim that the monodromy representation of the system satisfies the conditions of Theorem 6.1.1.

It follows from the construction of this system that conditions a) and c) of Theorem 6.1.1 are satisfied. Indeed, condition c) is fulfilled by the construction from Proposition 6.3.3. Reducibility of χ follows from the fact that the column vector $(1, 0, 0)$ is the solution of the system and from Lemma 4.2.1.

Each matrix $\tilde{B}^i$, $i \neq 1$, has single eigenvalue 0 and has rank two, therefore it can be transformed to one Jordan block. As in Proposition 6.3.3 by (6.3.12) we get that the same is true for the matrices E_i and for $G_i = \exp(2\pi i E_i)$.

Consider system (6.2.1), (6.5.2) in $\tilde{U}_1^*$. Let $Y_1(\tilde{x})$ be the fundamental matrix of (6.2.1), (6.3.11) given by (6.3.13). Then

$$Y = \begin{pmatrix} 1 & y_{12} & y_{13} \\ 0 & & \\ 0 & & Y_1 \end{pmatrix} \qquad (6.5.3)$$

is the fundamental matrix for the considered system. Substituting $Y(\tilde{x})$ into this system, we get

$$\begin{cases} \frac{dy_{12}}{dx} = x^{-\gamma'-1}y_{22} + \sum_{i=2}^n \frac{1}{x-a_i}(b_{12}^i y_{22} + b_{13}^i y_{32}), \\[2mm] \frac{dy_{13}}{dx} = x^{-\gamma'-1}y_{23} + \sum_{i=2}^n \frac{1}{x-a_i}(b_{12}^i y_{23} + b_{13}^i y_{33}) \end{cases} \qquad (6.5.4)$$

Since it follows from Remark 6.3.1 that

$$\varphi_1(y_{22}) = \gamma', \ \varphi_1(y_{23}) \geq 0, \ \varphi_1(y_{33}) = -\gamma', \ \varphi_1(y_{32}) \geq \gamma',$$

where $\varphi_1(f)$ is the valuation of f at $a_1 = 0$, we find from (6.5.4) that

$$\varphi_1\left(\frac{dy_{12}}{dx}\right) = -1, \ \varphi_1\left(\frac{dy_{13}}{dx}\right) = -\gamma' - 1. \qquad (6.5.5)$$

So we get that $a_1 = 0$ is a regular singular point for our system.

The matrix E_1 in factorization (2.2.45) of $Y(\tilde{x})$ has the form

$$E_1 = \begin{pmatrix} 0 & \alpha & \beta \\ 0 & 0 & 1 \\ 0 & 0 & 0 \end{pmatrix}. \qquad (6.5.6)$$

If the equality $\alpha = 0$ were satisfied, $\tilde{x}^{E_1}$ would be of the form

$$\tilde{x}^{E_1} = \begin{pmatrix} 1 & 0 & \beta \ln \tilde{x} \\ 0 & 1 & \ln \tilde{x} \\ 0 & 0 & 1 \end{pmatrix}$$

and y_{12} in (6.5.3) would be a single-valued meromorphic function in a neighborhood of $a_1 = 0$. But then $\varphi_1\left(\frac{dy_{12}}{dx}\right)$ would be non-negative if $y_{12}(\tilde{x})$ were holomorphic and it would be less than -1 if $y_{12}(\tilde{x})$ had a pole at $a_1 = 0$. We have obtained a contradiction with the first equality in (6.5.5). This means that $\alpha \neq 0$ in (6.5.6) and E_1 can be reduced to a Jordan block. Therefore, G_1 has the same property. Condition b) of Theorem 6.1.1 is satisfied.

Thus, Hilbert's 21st problem for the monodromy representation χ of system (6.2.1), (6.5.2) has a negative solution.

2. Let p be greater than three. Denote by $G_1, \ldots, G_n$ the monodromy matrices of the system, constructed above. We may assume that

$$G_i = \begin{pmatrix} 1 & * \\ 0 & \\ 0 & G_i' \end{pmatrix}, \quad i = 1, \ldots, n$$

and

$$G_n = \begin{pmatrix} 1 & 1 & 0 \\ 0 & 1 & 1 \\ 0 & 0 & 1 \end{pmatrix}.$$

Consider a representation χ of dimension $p > 3$ with the monodromy matrices

$$\chi(\sigma_i) = \begin{pmatrix} 1 & 1 & 0 & \cdots & & \cdot & 0 \\ & \ddots & & & & & \\ & & 1 & 1 & & & \\ & & 1 & 1 & 0 & 0 & \\ 0 & & & & G_i & \end{pmatrix}, \quad i = 1, \ldots, n-1,$$

$$\chi(\sigma_n) = \begin{pmatrix} 1 & (-n+1) & 0 & \cdots & & \cdot & 0 \\ & \ddots & & & & & \\ & & 1 & (-n+1) & & & \\ & & & 1 & (-n+1) & \alpha_1 & \alpha_2 \\ & & & & 1 & 1 & 0 \\ & & & & 0 & 1 & 1 \\ 0 & & & & 0 & 0 & 1 \end{pmatrix},$$

where α_1, α_2 are chosen in the following way. The product G of the matrices $\chi(\sigma_i), i \neq n$ is equal to

$$G = \begin{pmatrix} 1 & (n-1) & 0 & \cdots & & \cdot & 0 \\ & \ddots & & & & & \\ & & 1 & (n-1) & & & \\ & & & 1 & (n-1) & \beta_1 & \beta_2 \\ & & & & 1 & -1 & 0 \\ & & & & 0 & 1 & -1 \\ 0 & & & & 0 & 0 & 1 \end{pmatrix}.$$

We choose $\alpha_1 = -n+1-\beta_1$, $\alpha_2 = -\beta_1 - \beta_2$. Thus, we have that

$$\prod_{i=1}^{n} \chi(\sigma_i) = I$$

and each $\chi(\sigma_i)$ can be transformed to Jordan normal form consisting of only one block with eigenvalue 1. Therefore, we are under assumptions of Theorem 5.3.1. Thus, Hilbert's 21st problem for the monodromy representation with the singular points $a_1, \ldots, a_n$ and with the monodromy matrices $\chi(\sigma_1), \ldots, \chi(\sigma_n)$ has a negative solution.

It follows from Theorem 6.2.2 that the counterexample of Chapter 2 is the minimal possible (with respect to p and n).

From Theorem 6.4.1 and Theorem 6.1.1 it also follows that all counterexamples to Hilbert's 21st problem in dimension $p = 3$ are unstable in the following sense. If one slightly perturbs the singular points $a_1, \ldots, a_n$ without changing the monodromy matrices $\chi(\sigma_1), \ldots, \chi(\sigma_n)$, then the answer to Hilbert's 21st problem can become positive. Indeed, by Theorem 6.4.1 under some slight move of the singular points we obtain $\gamma_{\chi(a')} = 0$ or $\gamma_{\chi(a')} = 1$. Since by the proof of Theorem 6.2.2 $\gamma_{\chi(a)}$ is even, we get $\gamma_{\chi(a')} = 0$. Thus, by Theorem 6.1.1 the representation $\gamma_{\chi(a')}$ can be realized as the monodromy representation of some Fuchsian system. More precisely, in terms of Section 6.4 the following statement holds.

Theorem 6.5.2 *For any sequence of points $a = (a_1, \ldots, a_n)$, any representation χ of dimension $p = 3$, and any $\varepsilon > 0$, there is a sequence of points $a' = (a'_1, \ldots, a'_n)$ such that $|a'_i - a_i| < \varepsilon$ and Hilbert's 21st problem for $\chi(a')$ has a positive solution.*

This theorem means that there is no a sequence of $(3,3)$-matrices $G_1, \ldots, G_n$ with the condition $G_1 \cdot \ldots \cdot G_n = I$ such that Hilbert's 21st problem for the representation $\chi(a)$ has a negative solution for all sequences of points $a = (a_1, \ldots, a_n)$.

The following statement is a straightforward corollary of constructions of Theorem 6.5.1.

Corollary 6.5.1 *Let a representation χ of dimension $p = 3$ satisfy conditions a), b), c) of Theorem 6.1.1. Then there is a system (1.2.1) with the monodromy χ and with regular singular points $a_1, \ldots, a_n$ that is Fuchsian at all the points except one, where the order of the pole of the matrix of coefficients is equal to $\gamma_{\chi_2} + 1$, and there are no such systems with a pole of order less than $\gamma_{\chi_2} + 1$.*

From Proposition 6.3.2 and previous corollary we obtain the following result.

Corollary 6.5.2 *For any representation χ of dimension $p = 3$ there is a system (1.2.1) with regular singular points $a_1, \ldots, a_n$ and with monodromy χ that is Fuchsian at all the points except perhaps one, where the order of the pole of the matrix of coefficients does not exceed $\left[\frac{n}{2}\right]$, where $[x]$ is the integer part of x.*

7 Fuchsian equations

7.1 The number of apparent singularities

In introduction we claimed that for $p > 1$ a Fuchsian equation of pth order with singularities $a_1, \ldots, a_n$ contains fewer parameters than the set of classes of conjugate representations $\chi : \Delta \longrightarrow GL(p, \mathbb{C})$. This set depends on $N_r = p^2(n-2) + 1$ parameters (cf. (5.6.1)). Let us calculate the number of parameters for a Fuchsian equation of the pth order with n singular points $a_1, \ldots, a_n$.

We may assume that for all $i, a_i \neq \infty$. Then by (1.2.12) we obtain

$$q_i(x) = \frac{p_i(x)}{(x - a_1)^i \cdots (x - a_n)^i}, \quad i = 1, \ldots, n, \tag{7.1.1}$$

where $p_i(x)$ are analytic in $\mathbb{C}$. We must inquire when equation (1.2.11), (7.1.1) has no singularity at ∞. In order to do so we must rewrite (7.1.1) in terms of the new independent variable $\zeta = \frac{1}{x}$. It follows from (4.2.16) that under this change equation (1.2.11) is transformed to the following one:

$$\frac{d^p y}{d\zeta^p} + \tilde{q}_1(\zeta) \frac{d^{p-1} y}{d\zeta^{p-1}} + \cdots + \tilde{q}_p(\zeta) y = 0,$$

where

$$\tilde{q}_i(\zeta) = \frac{p_i(x) x^{2(p-1)}}{(x - a_1)^i \cdots (x - a_n)^i} \left(\frac{1}{x^{2(i-1)}} + o(\frac{1}{x^{2(i-1)}}) \right). \tag{7.1.2}$$

From the condition of holomorphy for $\tilde{q}_i(\zeta)$ at $\zeta = 0$ we get that $p_i(x)$ has at most a pole of the order n_i at $x = \infty$ and

$$n_i \leq (n + 2)i - 2p, \tag{7.1.3}$$

therefore it is a polynomial in x of degree n_i. Since any polynomial of degree n_i contains $n_i + 1$ parameters (which are the coefficients of the polynomial), we obtain that any Fuchsian equation (1.2.11) on $\bar{\mathbb{C}}$ with n singular points depends on

$$N_e = \sum_{i=1}^{p} (n_i + 1) = \frac{(n + 2)p(p + 1)}{2} - 2p^2 + p = \frac{(n - 2)p^2}{2} + \frac{pn}{2}$$

parameters.

The difference d^0 between N_r and N_e is equal to

$$d^0 = N_r - N_e = \frac{(n-2)p(p-1)}{2} + 1 - p \qquad (7.1.4)$$

This difference is greater than 0 for all $n > 3, p > 1$ (or $n = 3, p > 2$), therefore in general a construction of a Fuchsian equation with a given monodromy is possible only in the case of an appearance of additional so-called apparent singularities. (Recall that these singularities are called *apparent*, because they are not ramification points for solutions of a equation. Let us estimate the number of such singularities . Consider a vector bundle G^λ, constructed by a representation χ and by some admissible set $(\lambda^1, \ldots, \lambda^n)$ (cf. Section 5.1).

Theorem 7.1.1 *For any irreducible representation χ and admissible set λ there exists a Fuchsian equation (1.2.11) with the given monodromy, which has a number m of additional apparent singular points, satisfying the following inequality*

$$m \leq \frac{(n-2)p(p-1)}{2} - \gamma(\lambda) + 1 - l, \qquad (7.1.5)$$

where $\gamma(\lambda)$ is the weight of G^λ, l is a number of terms c_i in decomposition (5.1.6), which are equal to c_1. ([Bo3].)

Proof. The proof consists in a small modification of the proof of Proposition 4.2.1. Consider a meromorphic holomorphically invertible off a_1 section of the bundle R^λ and consider the corresponding fundamental matrix $Y(\tilde{x})$ from (5.1.2), constructed by the section. Let $c_1 = \cdots = c_l \neq c_{l+1}$ in factorizations (4.2.1), (4.2.2) for the matrix $\tilde{Y}(\tilde{x}) = \Gamma_1 Y(\tilde{x})$ with $B = \text{diag}(\lambda_1^1, \ldots, \lambda_p^1)$. Then there exists a nondegenerate matrix

$$S = \left(\begin{array}{cc} S' & 0 \\ 0 & I^{p-l} \end{array} \right)$$

such that the first row of the matrix $S(x - a_1)^C V(x)$ has the form

$$(\underbrace{0, \ldots, 0}_{l-1}, t_1, \ldots, t_{p-l+1})$$

at a_1. Denote by $y_1, \ldots, y_p$ the first row of the matrix $S\tilde{Y}(\tilde{x})$. Then, for this row factorizations (4.2.10) hold with replacing $B + c_1 I$ by $B + c_1 I + F$, where $F = \text{diag}(\underbrace{1, \ldots, 1}_{l-1}, 0, \ldots, 0)$. Thus, we also may replace $B + c_1 I$ by $B + c_1 I + F$.

By the same way as in the proof of Proposition 4.2.1 , we get

$$pc_1 + l - 1 + \text{tr}B + \sum_{i=2}^{n} \text{tr}\Phi_i + \sum_{i=1}^{n} \text{tr}E_i + d = \frac{(n-2)p(p-1)}{2} \qquad (7.1.6)$$

instead of (4.2.20). Subtracting (4.2.9) from (7.1.6), we obtain

$$d = \sum_{i=1}^{p}(c_1 - c_p) + l - 1 = \frac{(n-2)p(p-1)}{2}. \qquad (7.1.7)$$

Consider the equation

$$\frac{1}{W(\tilde{x})} \begin{vmatrix} y & y_1 & \cdots & y_p \\ \frac{dy}{dx} & \frac{dy_1}{dx} & \cdots & \frac{dy_p}{dx} \\ \vdots & \vdots & \cdots & \vdots \\ \frac{d^p y}{dx^p} & \frac{d^p y_1}{dx^p} & \cdots & \frac{d^p y_p}{dx^p} \end{vmatrix} = 0 \qquad (7.1.8)$$

This equation is Fuchsian at the points $a_1, \ldots, a_n$ and has the given monodromy. Its additional singularities are the points $\tilde{b}_1, \ldots, \tilde{b}_m$, which are zeroes of the Wronskian $W(\tilde{x})$. Obviously, these points and the orders d_j of zeroes of $W(\tilde{x})$ coincide with that for $\det T'(\tilde{x})$ from (4.2.12).

Since by the formula below (4.2.20)

$$d = \sum_{i=1}^{m} d_i + s,$$

where $d_1, \ldots, d_m$ are the orders of the zeroes for $W(\tilde{x})$ at $\tilde{b}_1, \ldots, \tilde{b}_m$ and since $s \geq 0$, we have that $m \leq d$. Thus, from (7.1.7) we get (7.1.5).

Remark 7.1.1 *Formula (7.1.5) improves the estimate*

$$m \leq \frac{(n-2)p(p-1)}{2} + 1 - p,$$

which follows from the corresponding inequality in [Oht], obtained there for any compact Riemann surface (under the assumption that one of the monodromy matrices is semisimple). In the case $c_1 - c_p \leq 1$ both the estimates coincide.

Example 7.1.1 *Consider system (1.2.1), (2.1.4) . Since the weight $\gamma(0)$ of the canonical extension G^0, constructed by the monodromy χ of the system, is equal to two (see Section 2.3 and Theorem 6.2.1), we have $\gamma(0) = 0, l = 1$, and $m = 0$ in (7.1.5). Thus, the monodromy representation of this system can be realized as the monodromy representation of some Fuchsian equation (1.2.11) with the singular points $a_1, \ldots, a_n$ without additional singularities.*

7.2 Fuchsian equations and systems

The main result of the part is formulated as follows

Theorem 7.2.1 *For any Fuchsian equation (1.2.11) on the Riemann sphere there always exists a Fuchsian system with the same singular points and the same monodromy.*

Proof. In order to prove the theorem we need to extend some constructions from local theory of Fuchsian systems on Fuchsian equations.

The concepts of a Levelt's filtration and of a Levelt's basis for equations are the same as for systems. If $y_1, \ldots, y_p$ is a Levelt's basis in $\tilde{U}_i^*$ for (1.2.11), then instead of (2.2.25) we have the factorization

$$(y_1, \ldots, y_p) = (v_1, \ldots, v_p)(x - a_i)^{\Phi_i}(\tilde{x} - a_i)^{E_i} \tag{7.2.1}$$

with holomorphic v_j and with $v_1(a_i) \neq 0$.

Lemma 7.2.1 *If in factorization (7.2.1) for a Levelt's basis $y_1, \ldots, y_p$ the equality $v_j(a_i) = 0$ holds, then*

$$\varphi_i^{j-1} = \varphi_i^j.$$

Proof. From the fact that E_i is upper-triangular it follows that

$$(y_1, \ldots, y_j) = (v_1, \ldots, v_j)(x - a_i)^{\Phi_i^j}(\tilde{x} - a_i)^{E_i^j}, \tag{7.2.2}$$

where $\Phi_i^j = \mathrm{diag}(\varphi_i^1, \ldots, \varphi_i^j)$, E_i^j is formed by intersections of the first j rows and columns of E_i. Therefore,

$$\left(\frac{y_1}{(x - a_i)^{\varphi_i^j + 1}}, \ldots, \frac{y_j}{(x - a_i)^{\varphi_i^j + 1}} \right) =$$

$$= \left(\frac{v_1}{x - a_i}, \ldots, \frac{v_j}{x - a_i} \right)(x - a_i)^{\Phi_i^j - \varphi_i^j I}(\tilde{x} - a_i)^{E_i^j}.$$

If $v_j(a_i) = 0, \varphi_i^{j-1} > \varphi_i^j$, then all $\frac{v_l}{x - a_i}, l = 1, \ldots, j$ are holomorphic at a_i and the elements of $\Phi_i^j - \varphi_i^j I$ are nonnegative. Thus, from the definition of valuations we get the inequality

$$\varphi_i(y_j) \geq \varphi_i^j + 1$$

which contradicts the choice of the basis $y_1, \ldots, y_p$. Therefore, $\varphi_i^{j-1} = \varphi_i^j$.

The exponents $\beta_i^1, \ldots, \beta_i^p$ are the roots of *the indical equation*

$$\lambda(\lambda - 1) \cdots (\lambda - p + 1) + \lambda(\lambda - 1) \cdots (\lambda - p + 2)r_1(a_i) + \cdots +$$

$$+ \lambda(\lambda - 1) \cdots (\lambda - p + j + 1)r_j(a_i) + \cdots + r_p(a_i) = 0 \qquad (7.2.3)$$

for equation (1.2.11).

Indeed, let $y_1, \ldots, y_p$ be a strongly Levelt's basis (see Section 2.2) and let for a given β_i^j we have that either $j = 1$ or $\varphi_i^{j-1} > \varphi_i^j$. Then by (7.2.1) we obtain

$$y_j = (\tilde{x} - a_i)^{\beta_i^j} v_j(x)(1 + o((\tilde{x} - a_i)^{1/2} \ln^{j-1}(\tilde{x} - a_i))),$$

where $v_j(x)$ is holomorphic at a_i. Substituting y_j in (1.2.11) and multiplying the result by $(\tilde{x} - a_i)^{-\beta_i^j}$, for $x = a_i$ we get equality (7.2.3) with $\lambda = \beta_i^j$.

Inversely, if λ_0 is a root of (7.2.3), then there exists a solution of (1.2.11) of the form $y = (\tilde{x} - a_i)^{\lambda_0} h(x)$ with holomorphic $h(x), h(a_i) \neq 0$. It follows from (7.2.1), that λ_0 is an exponent β_i^j for some j (see [In] for details).

Lemma 7.2.2 *Let the function $h(x)$ be holomorphic in the neighborhood U_l of the point a_l and $h(a_l) \neq 0$. System (1.2.1), constructed from equation (1.2.11) by means of the substitution*

$$f^j(\tilde{x}) = ((x - a_l)h(x))^{j-1} \frac{d^{j-1}y}{dx^{j-1}}, \qquad (7.2.4)$$

is Fuchsian at the point a_l and its exponents β_l^j at that point coincide with those of equation (1.2.11).

Proof. A straightforward computation shows that the matrix A of the system (1.2.1) constructed by means of the substitution (7.2.4) from equation (1.2.11), (1.2.12) has the following form: $A = \frac{1}{x - a_l} B$, where $B =$

$$\begin{pmatrix} 0 & \frac{1}{h} & 0 & \cdots & \cdots & \cdots & 0 \\ 0 & \frac{((x-a_l)h)'}{h} & \frac{1}{h} & 0 & \cdots & \cdots & 0 \\ \vdots & \vdots & \vdots & \cdots & \cdots & \cdots & \vdots \\ 0 & \cdots & \cdots & \frac{(j-1)((x-a_l)h)'}{h} & \frac{1}{h} & \cdots & 0 \\ \vdots & \vdots & \vdots & \cdots & \cdots & \cdots & \vdots \\ -r_p h^{p-1} & \cdots & \cdots & \cdots & \cdots & -r_2 h & -\frac{(r_1 - (p-1))((x-a_l)h)'}{h} \end{pmatrix},$$

therefore the constructed system is Fuchsian at the point a_l.

By (7.2.4) and the definition of valuations it follows that

$$\varphi_i(f^j) \geq \varphi_i(y),$$

thus, for any solution f of the constructed system we have

$$\varphi_i(f) = \varphi_i(y) = \varphi_i^j,$$

since $f^1 = y$. Let $y_1, \ldots, y_p$ be a Levelt's basis in $\tilde{U}_i^*$ for (1.2.11). The latter equalities mean that the basis $f_1, \ldots, f_p$, constructed from $y_1, \ldots, y_p$ by (7.2.4) is a Levelt's basis for the system. Thus, exponents of the system coincide with that of equation.

Consider again the proof of Proposition 4.2.1 and apply it to the row $y_1, \ldots, y_p$, where $y_1, \ldots, y_p$ is a Levelt's basis for Fuchsian equation (1.2.11). Since Wronskian of these functions has no additional zeroes or poles, we have that $d_j = 0$ in (4.2.12) and $d' = 0$ in (4.2.19). From the fact that the constructed system is Fuchsian at a_l we get that $\det V_l'(a_l) \neq 0$, therefore $s_i = 0$ in (4.2.15). Thus $d = 0$ in (4.2.20) and of course $c_1 = 0$, $B = 0$ (we started from $y_1, \ldots, y_p$ with $c_1 = 0$ and with any B). So, from (4.2.20) we get the classical *Fuchsian relation*

$$\sum_{i=1}^{n} \sum_{j=1}^{p} \beta_i^j = \frac{(n-2)p(p-1)}{2} \tag{7.2.5}$$

for Fuchsian equation (1.2.11) on the Riemann sphere.

Consider factorization (2.2.25) for system (1.2.1), constructed from (1.2.11) by (7.2.4). Denoting by $T_l(\tilde{x})$ the corresponding Levelt's matrix, we obtain

$$T_l(\tilde{x}) = V_l(x)(x - a_l)^{\Phi_l}(\tilde{x} - a_l)^{E_l}, \quad \tilde{x} \in \tilde{U}_l^*. \tag{7.2.6}$$

Lemma 7.2.3 *All principal minors of the matrix $V_l(x)$ in factorization (7.2.6) for the fundamental matrix $T_l(\tilde{x})$ are different from zero at the point a_l.*

Proof. Consider the first components $y_1, \ldots, y_n$ of the first i columns of the matrix T_l. The analytic functions $y_1(\tilde{x}), \ldots, y_i(\tilde{x})$ are linearly independent, and the space X_i that they span is invariant under the action of the monodromy operator σ_l^* (this follows from the upper-triangularity of the matrix E_l in decomposition (7.2.6)). Therefore, X_i is the space of solutions of some equation (7.1.8) that is Fuchsian in U_l. From the uniqueness theorem for the Wronskian $W_i(\tilde{x})$ of the functions $y_1, \ldots, y_i$ it follows that the number of additional apparent singularities of this equation in U_l is finite. Let $O_l \subset U_l$ be such that O_l does not contain those points. Now the assertion of the lemma follows from the fact that the ith principal minor of the matrix $V_l(x)$ in question coincides with the determinant $\det V_l^i(x)$ of the matrix $V^i(x)$ figuring in decomposition (7.2.6) for the fundamental matrix $T^i(\tilde{x})$, constructed from the functions $y_1(\tilde{x}), \ldots, y_i(\tilde{x})$ by means of the substitution (7.2.4), and , by Lemma (7.2.2) this last determinant does not vanish at the point a_l.

Proof of Theorem 7.2.1. With no loss of generality one can assume that $a_1 = 0$ and ∞ is not a singular point of equation (1.2.11) (this can be always achieved by a suitable conformal mapping of the Riemann sphere). Use substitution (4.2.11)

$$f^j = \left(\prod_{i=1}^{n} (x - a_i)^{j-1} \right) \frac{d^{j-1}y}{dx^{j-1}} \qquad (7.2.7)$$

to pass from Fuchsian equation (1.2.11) to system (1.2.1). By Lemma 7.2.2 we have that this system is Fuchsian at the points $a_1, \ldots, a_n$, and its additional apparent singular point ∞ is a regular singular point. The fundamental matrix $T(\tilde{x})$ of the space of solutions of the system admits a decomposition

$$T(\tilde{x}) = \Gamma_2(x)\Gamma_1(x)R\left(\frac{1}{x}\right), \qquad (7.2.8)$$

where $R(\frac{1}{x})$ is from (4.2.17), $\Gamma_2\Gamma_1 = \tilde{\Gamma}_2$ in (4.2.17), $\Gamma_1(x) = (\gamma_{ij})$,

$$\gamma_{11} = 1, \ \gamma_{j1} = 0, \ j \neq 1, \ \gamma_{ji} = k_i^j x^{j-1}, \ 1 < i \leq j, \qquad (7.2.9)$$

and

$$\Gamma_2 = \text{diag}\left(1, \frac{1}{x^2} \prod_{l=1}^{n} (x - a_l), \ldots, \frac{1}{x^{2(p-1)}} \prod_{l=1}^{n} (x - a_l)^{p-1}\right)$$

(such a factorization follows from (4.2.16)).

Denote by $\Gamma_3(x), \Gamma_4(x)$ the matrices

$$\Gamma_3(x) = \text{diag}(1, x^{-(n-2)}, \ldots, x^{-(n-2)(p-1)}),$$

$$\Gamma_4(x) = x^{(n-2)(p-1)}\Gamma_3(x) = x^C, \qquad (7.2.10)$$

where $C = \text{diag}((n-2)(p-1), \ldots, 0)$, and denote by $\Gamma_5(x)$ the matrix

$$\Gamma_5(x) = \Gamma_3\Gamma_2\Gamma_1\Gamma_2^{-1}\Gamma_3^{-1}.$$

From form (7.2.9) of the matrix $\Gamma_1(x)$–in particular, from its lower-triangularity–it follows that $\Gamma_5(x)$ is a lower-triangular matrix that is holomorphically invertible off the points $a_1 = 0$ and ∞ and such that the elements on its principal diagonal are nonzero constant numbers (equal to the numbers k_j^j from (4.2.16)). Hence, the same holds true for the matrix Γ_5^{-1}.

Lemma 7.2.4 *There exists a lower-triangular matrix $\Gamma_6(x)$, meromorphic on $\bar{\mathbb{C}}$ and holomorphically invertible off the point $a_1 = 0$, such that the matrix $\Gamma_7(x) = \Gamma_6(x)\Gamma_5^{-1}(x)$ is holomorphically invertible off the point ∞ and is lower-triangular with all elements on the principal diagonal equal to 1.*

Proof. Let $\Gamma_5^{-1}(x) = (\gamma_{ji})$, $S = \text{diag}(\gamma_{11}, \ldots, \gamma_{pp})$. The matrix $\Gamma_7^0(x) = S^{-1}\Gamma_5^{-1}(x)$ has unit principal diagonal and is meromorphic at the point 0. Each element γ_{ji}^0 of Γ_7^0 has the form $\gamma_{ji}^0(x) = Q_0^{ji}(1/x) + h_{ji}^0(x)$, where $Q_0^{ji}(1/x)$ is a polynomial and the function $h_{ji}^0(x)$ is holomorphic at zero. Consequently, the elements in the $(p-1)$st and pth column of the matrix $\Gamma_7^1 = \Gamma_6^{p-1} \cdot \Gamma_7^0$, where

$$\Gamma_6^{p-1} = \begin{pmatrix} 1 & & & 0 \\ & \ddots & & \\ & & 1 & \\ 0 & & -Q_0^{pp-1} & 1 \end{pmatrix}$$

are holomorphic at zero. The remaining elements γ_{ji}^1 of this matrix are also of the form $\gamma_{ji}^1(x) = Q_1^{ji}(1/x) + h_{ji}^1(x)$. For the matrix $\Gamma_7^2 = \Gamma_6^{p-2} \cdot \Gamma_7^1$, where

$$\Gamma_6^{p-2} = \begin{pmatrix} 1 & & & & 0 \\ & \ddots & & & \\ & & 1 & & \\ & & -Q_1^{p-1p-2} & 1 & 0 \\ 0 & & -Q_1^{pp-2} & 0 & 1 \end{pmatrix},$$

already the $(p-2)$th and $(p-1)$st columns are holomorphic, and so on. Iterating the procedure whose first two steps were just described, after $p-2$ steps we obtain the sought-for matrices $\Gamma_6 = \left(\prod_{i=2}^{p-1} \Gamma_6^i\right) S^{-1}$ and $\Gamma_7(x) = \Gamma_6(x)\Gamma_5^{-1}(x)$. The lemma is proved.

The matrix $\Gamma_7(x)\Gamma_4(x)$ can be written in the form

$$\Gamma_7(x)\Gamma_4(x) = \Gamma_4(x)\Gamma_8(x), \tag{7.2.11}$$

where $\Gamma_8(x) = \Gamma_4^{-1}(x)\Gamma_7(x)\Gamma_4(x)$ is a lower-triangular matrix that is holomorphically invertible off the point ∞, meromorphic at ∞, and with the elements on the principal diagonal equal to 1.

Next, let us pass from the system (1.2.1), constructed by means of substitution (7.2.7) with the fundamental matrix T to the system (1.2.1) with the fundamental matrix

$$T' = \Gamma_4(x)\Gamma_8(x)x^{-(n-2)(p-1)}T.$$

Lemma 7.2.5 *The system (1.2.1) with the fundamental matrix T' is Fuchsian at the points $a_2, \ldots, a_n$, and has $a_1 = 0$ as a regular singular point and ∞ as a point of holomorphy.*

Proof. That the constructed system is Fuchsian at the points $a_2, \ldots, a_n$ follows from the fact that $\det(\Gamma_4(a_l)\Gamma_8(a_l)) \cdot a_l^{-(n-2)(p-1)} \neq 0$ and from holomorphy of $\Gamma_4\Gamma_8$ at a_l. From relations (7.2.9) and (7.2.9)–(7.2.11) we obtain

$$\Gamma_4\Gamma_8 x^{-(n-2)(p-1)} T(\tilde{x}) = \Gamma_7\Gamma_3 T(\tilde{x}) = \Gamma_6\Gamma_5^{-1}\Gamma_3 T(\tilde{x}) =$$

$$\Gamma_6\Gamma_3\Gamma_2\Gamma_1^{-1}\Gamma_2^{-1}\Gamma_3^{-1}\Gamma_3 T(\tilde{x}) = \Gamma_6\Gamma_3\Gamma_2 R(\frac{1}{\tilde{x}}).$$

Since the matrices $\Gamma_6(x)$, $\Gamma_3(x) \cdot \Gamma_2(x)$ are holomorphically invertible at ∞ and $R(1/\tilde{x})$ is holomorphically invertible at the points $p^{-1}(\infty)$ of the universal covering $\tilde{S}$, we get that ∞ is not singular for the constructed system. The lemma is proved.

Factorization (7.2.6) for T' at the point $a_1 = 0$ is connected with a similar factorization for T as follows

$$T'(\tilde{x}) = \Gamma_4(x)V_1^0(x)x^{\Phi_1'}\tilde{x}^{E_1}, \tag{7.2.12}$$

where $V_1^0(x) = \Gamma_8(x)V_1(x)$, $\Phi_1' = \Phi_1 - (n-2)(p-1)I$. Since the matrix $\Gamma_8(x)$ is lower-triangular and holomorphically invertible at zero, Lemma 7.2.3 guarantees that all principal minors of the matrix $V_1^0(x)$ are different from zero at $a_1 = 0$. Next, it follows from the form of the matrix Φ_1' that valuations $\tilde{\varphi}_1^j$ of T' at zero are connected with the corresponding valuations φ_1^j of T by the relations

$$\tilde{\varphi}_1^j \geq \varphi_1^j - (n-2)(j-1), \quad j = 1, \ldots, p. \tag{7.2.13}$$

By Lemma 4.1.2 there exists a meromorphic on $\bar{\mathbb{C}}$ holomorphically invertible off $a_1 = 0$ matrix $\Gamma(x)$ such that

$$\Gamma(x)\Gamma_4(x)V_1^0(x)x^{\Phi_1'}\tilde{x}^{E_1} = \tilde{V}_1(x)x^{\Phi_1'+C}\tilde{x}^{E_1} \tag{7.2.14}$$

with C from (7.2.10) and with $\det \tilde{V}_1(0) \neq 0$. Consider the system with the fundamental matrix $\tilde{T}(\tilde{x}) = \Gamma(x)T'(\tilde{x})$. This system is already Fuchsian at all the points $a_1, \ldots, a_n$. The theorem is proved.

Let us again denote by $\tilde{\beta}_i^j$ the exponents of the constructed system. From (7.2.13), (7.2.14) and (7.2.10) we obtain

Corollary 7.2.1 *The exponents β_i^j of the original Fuchsian equation (1.2.11) and the exponents $\tilde{\beta}_i^j$ of the Fuchsian system (1.2.1), constructed according to Theorem 7.2.1 are connected by the equalities*

$$\tilde{\beta}_1^j = \tilde{\beta}_1^j - (n-2)(j-1), \quad \tilde{\beta}_i^j = \tilde{\beta}_i^j, \tag{7.2.15}$$

$$i \neq 1, \quad j = 1, \ldots, p.$$

From Theorem 7.2.1 we also get the following sequence of statements.

Corollary 7.2.2 *The solutions of the original Fuchsian equation (1.2.11) coincide with the first components of the system (1.2.1) constructed in Theorem 7.2.1.*

Proof immediately follows from the forms of transformation (7.2.7) and of the transformations used in the theorem .

Proposition 7.2.1 *Suppose the representation χ is reducible and each monodromy matrix $G_i = \chi(\sigma_i)$ can be reduced to a Jordan block. Then there is no Fuchsian equation (1.2.11) without additional apparent singular points, the monodromy of which coincides with χ.*

Proof. Let such a Fuchsian equation exist. Then by Theorem 7.2.1 and Corollary 7.2.1 there exists a Fuchsian system with the same monodromy and with valuations $\varphi_i^1, \ldots, \varphi_i^p$, such that $\varphi_i^1 > \varphi_i^2$. But this inequality contradicts the statement of Proposition 5.2.1. Thus, such an equation does not exist.

Proposition 7.2.1 can be directly deduced from Fuchsian relation (7.2.5) as follows.

If the matrix $\chi(\sigma_i)$ can be reduced to one Jordan block, then by the properties (2.2.6) and (2.2.8) of valuations one has

$$\varphi_i^1 \geq \cdots \geq \varphi_i^p \tag{7.2.16}$$

for a Fuchsian equation at a_i. Indeed, the Jordan basis $e_1, \ldots, e_p$ is also the Levelt basis in this case (cf. the proof of Proposition 5.2.1). And we obtain that

$$\varphi_i^j = \varphi_i(e_j) = \varphi_i(\sigma_i^* e_j) = \varphi_i(e_j + e_{j-1}) \geq \min\{\varphi_i^j, \varphi_i(e_{j-1})\},$$

therefore by (2.2.8): $\varphi_i(e_{j-1}) = \varphi_i^{j-1} \geq \varphi_i^j$.

If χ is reducible, then for some l, $1 \leq l < p$, l independent solutions $y_1, \ldots, y_l$ of the equation generate Fuchsian equation (7.1.8) (with $p = l$). Since for each singular point a_i : $(y_1, \ldots, y_l) = \mathbb{C}(e_1^i, \ldots, e_l^i)$, where $e_1^i, \ldots, e_l^i$ are the first l elements of the corresponding Jordan basis at a_i, we get that the exponents β_i^j of the equation at $a_1, \ldots, a_n$ coincide of that for the original equation. The constructed equation may have additional apparent singularities $b_1, \ldots, b_m$, where $\tilde{\varphi}_k^j = \varphi_k(y_j) \geq 0$. Thus, by (7.1.8) we get

$$\sum_{i=1}^{n} \sum_{j=1}^{l} \beta_i^j + \sum_{k=1}^{m} \sum_{j=1}^{l} \tilde{\varphi}_k^j = \frac{(n+m-2)l(l-1)}{2}. \tag{7.2.17}$$

Let us show that

$$\sum_{j=1}^{l} \tilde{\varphi}_k^j \geq \frac{ml(l-1)}{2} \tag{7.2.18}$$

for all $k = 1, \ldots, m$. Consider a basis $y_1, \ldots, y_l$ in the space of solutions for the constructed equation such that

$$y_1(\tilde{x}) = (x - b_k)^t h_1(\tilde{x}), \ldots, y_l(\tilde{x}) = (x - b_k)^{t+l-1} h_l(\tilde{x}), \tag{7.2.19}$$

where $h_j(\tilde{x})$, $j = 1, \ldots, l$ are holomorphic at $p^{-1}(b_k)$; $t \geq 0$. Obviously, such a basis always exists and can be constructed by induction. Indeed, if $y_j(\tilde{x}) = (x - b_k)^r h_j(\tilde{x})$, $j = 1, 2$, then a suitable linear combination $y_2' = s_1 y_1 + s_2 y_2$ has the form $(x - b_k)^{r+1} h_2'(\tilde{x})$, and so on. Thus, we get from (7.2.19) that $\varphi_k(y_j) \geq j - 1$, which implies (7.2.18).

Subtracting (7.2.18) from (7.2.17), we obtain

$$\sum_{i=1}^{n} \sum_{j=1}^{l} \beta_i^j \leq \frac{(n-2)l(l-1)}{2}. \tag{7.2.20}$$

But from inequalities $\varphi_i^t \geq \varphi_i^j$, $1 \leq t \leq l$, $l < j \leq p$ it follows the inequality

$$\sum_{i=1}^{n} \sum_{j=1}^{p} \beta_i^j \leq \frac{p}{l} \sum_{i=1}^{n} \sum_{j=1}^{l} \beta_i^j \leq$$

$$\leq \frac{p}{l} \frac{(n-2)l(l-1)}{2} = \frac{(n-2)p(l-1)}{2} < \frac{(n-2)p(p-1)}{2},$$

which contradicts Fuchsian relation (7.2.5).

Theorem 7.2.2 *An irreducible monodromy representation χ can be realized as the monodromy representation of some Fuchsian equation (1.2.11) without additional apparent singularities if and only if the maximal Fuchsian weight of χ is equal to $\frac{(n-2)p(p-1)}{2}$.*

Proof. If $\gamma_m(\chi) = \frac{(n-2)p(p-1)}{2}$, then from (7.1.5), applied to $\gamma_m(\chi)$ we get $m = 0$ for the number of apparent singularities of the Fuchsian equation, constructed in Theorem 7.1.1.

Let such an equation exist and let β_i^j, φ_i^j be its exponents and valuations respectively. Consider the bundle G^λ with $\lambda = (\lambda^1, \ldots, \lambda^n)$, where

$$\lambda^j = (\varphi_j^1, \ldots, \varphi_j^p), \ j = 2, \ldots, p,$$

$$\lambda^1 = (\varphi_1^1 + (n-2)(p-1), \ldots, \varphi_1^p).$$

But from (7.2.12) and (7.2.10) it follows that

$$G^\lambda \cong \mathcal{O}((n-2)(p-1)) \oplus \mathcal{O}((n-2)(p-2)) \oplus \cdots \oplus \mathcal{O}(0). \qquad (7.2.21)$$

(See Section 5.2 for details.) Thus,

$$\gamma(\lambda) = p(n-2)(p-1) - \sum_{j=1}^{p}(n-2)(p-j) = \frac{(n-2)p(p-1)}{2}.$$

Due to (7.1.5) we again obtain

$$\gamma_m(\chi) = \frac{(n-2)p(p-1)}{2}.$$

7.3 Examples

In what follows many calculations and other intermediate steps are left to the reader as exercises.

1. Consider the case $p = 2, n = 1$. The only Fuchsian equation (1.2.11) with at most one singularity $a_1 = 0$ is

$$\frac{d^2y}{dx^2} + \frac{2}{x}\frac{dy}{dx} = 0. \qquad (7.3.1)$$

(As we see, it really has a pole; so there are no Fuchsian equations (1.2.11) which have no singularities at all). Its solutions are $y = c_1 + c_2/x$; they do not branch. But $S = \bar{\mathbb{C}} \setminus \{0\}$ is simply connected, $\tilde{S} = S$ and $\Delta = \{\text{identity}\}$. So the monodromy must be trivial, and it is realizable by (7.3.1).

If $n = 2$, it is convenient to choose singular points $a_1 = 0$, $a_2 = \infty$. (This can be achieved using an appropriate conformal map of $\bar{\mathbb{C}}$). All Fuchsian equations (1.2.11) of the second order with these singularities are

$$\frac{d^2 y}{dx^2} + \frac{a}{x}\frac{dy}{dx} + \frac{b}{x^2}y = 0 \qquad (7.3.2)$$

with arbitrary constants a, b. This is well known Euler's equation; it can be solved easily. Introducing a new independent variable $t = \ln x$ we get

$$\frac{d^2 y}{dt^2} + (a - 1)\frac{dy}{dt} + by = 0.$$

If the indical equation

$$\lambda^2 + (a - 1)\lambda + b = 0 \qquad (7.3.3)$$

has two different roots λ_1, λ_2, then solutions to (7.3.2) are $y = c_1 x^{\lambda_1} + c_2 x^{\lambda_2}$. If (7.3.3) has a double root λ, then solutions to (7.3.2) are $y = c_1 x^{\lambda} \ln x + c_2 x^{\lambda}$. S is the "ring" $\mathbb{C} \setminus \{0\}$, $\Delta = \pi_1(S) = \mathbb{Z} = \{\sigma^k\}$, and we want to realize just the one matrix $G = \chi(\sigma)$. Take $y_1 = x^{\lambda_1}, y_2 = x^{\lambda_2}$ or $y_1 = x^{\lambda} \ln x, y_2 = x^{\lambda}$ for the basis in the space of solutions. Then for the deck transformation σ corresponding to one turn around the origin in the positive direction we obtain the monodromy matrices

$$\chi(\sigma) = \begin{pmatrix} e^{2\pi i \lambda_1} & 0 \\ 0 & e^{2\pi i \lambda_2} \end{pmatrix} \text{ or } \chi(\sigma) = \begin{pmatrix} e^{2\pi i \lambda} & 2\pi i e^{2\pi i \lambda} \\ 0 & e^{2\pi i \lambda} \end{pmatrix}. \qquad (7.3.4)$$

Clearly any matrix G is conjugate to one of the matrices (7.3.4). So in this case Hilbert's problem for equations has a positive solution.

The case $p = 2, n = 3$ (the case of the Riemann equation) is the unique nontrivial case for which the difference d^0 from (7.1.4) is equal to zero.

Proposition 7.3.1 *Each irreducible representation χ with $n = 3, p = 2$ can be realized as the monodromy representation of some Riemann equation. Representations which are conjugate to representations of the following form:*

$$\chi(\sigma_1) = \begin{pmatrix} \lambda_1 & c_1 \\ 0 & \lambda_1 \end{pmatrix}, \; \chi(\sigma_2) = \begin{pmatrix} \lambda_2 & c_2 \\ 0 & \lambda_2 \end{pmatrix},$$

$$\chi(\sigma_3) = \begin{pmatrix} \frac{1}{\lambda_1 \lambda_2} & c_3 \\ 0 & \frac{1}{\lambda_1 \lambda_2} \end{pmatrix}, \tag{7.3.5}$$

where $c_1 c_2 \neq 0$, $c_3 = -\frac{\lambda_1 c_2 + \lambda_2 c_1}{(\lambda_1 \lambda_2)^2} \neq 0$ *cannot be realized as the monodromy representations of any Riemann equations.*

Proof. It follows from Theorem 7.1.1 that $m = 0$ in (7.1.5), therefore each irreducible representation is realizable by a Riemann equation. From Theorem 7.2.2 it follows that each representation of form (7.3.5) is not realizable in this way.

The complete answer for a Riemann equation is presented in [KS]. It is proved there that all two-dimensional representations with three singular points can be realized as monodromy representations of Riemann equations except representations which are conjugated to representations of form (7.3.5) and of the next form:

$$\tilde{G}_1 = \begin{pmatrix} \lambda_1 & 0 \\ 0 & \lambda_2 \end{pmatrix}, \; \tilde{G}_2 = \begin{pmatrix} \mu_1 & 0 \\ 0 & \mu_2 \end{pmatrix}, \; \tilde{G}_3 = \tilde{G}_2^{-1}\tilde{G}_1^{-1},$$

with $\lambda_1 \neq \lambda_2$, $\mu_1 \neq \mu_2$, $\lambda_1 \mu_1 \neq \lambda_2 \mu_2$.

The constructions of Theorem 7.2.1, applied to the hypergeometric equation

$$\frac{d^2 y}{dx^2} + \frac{\gamma - (\alpha + \beta + 1)x}{x(1-x)} \frac{dy}{dx} - \frac{\alpha \beta}{x(1-x)} y = 0 \tag{7.3.6}$$

lead to the Fuchsian system

$$\frac{d\vec{y}}{dx} = \left(\begin{pmatrix} 0 & 0 \\ -\alpha\beta & -\gamma \end{pmatrix} \frac{1}{x} + \begin{pmatrix} 0 & 1 \\ 0 & \gamma - (\alpha + \beta) \end{pmatrix} \frac{1}{x-1} \right) \vec{y}.$$

(In this section, dealing simultaneously with vectors and scalars, we use arrows to indicate vectors).

2. Now as an illustration to local theory, presented in Section 2.2 we shall apply the local theory to one of the most well-known equations of the mathematical physics – Bessel's equation.

Let us first recall some basic facts about the equation

$$\frac{d^2 y}{dx^2} + \frac{p(x)}{x} \frac{dy}{dx} + \frac{q(x)}{x^2} y = 0 \tag{7.3.7}$$

with p, q holomorphic in some disk with the center at 0. This equation has a Fuchsian singularity at 0. The indical equation to (7.3.7) at 0 is as follows:

$$\lambda(\lambda - 1) + p(0)\lambda + q(0) = 0. \tag{7.3.8}$$

A short (but formal) way to derive (7.3.8) – or, rather, to remember it – is to replace in (7.3.7) $p(x)$ and $q(x)$ by their values at $x = 0$. The equation thus obtained is Euler's equation (7.3.2) with $a = p(0)$, $b = q(0)$. And (7.3.8) coincides with (7.3.3). Later we shall arrive to (7.3.8) in a less formal way tieing it with our previous considerations.

Let λ_1, λ_2 be roots of (7.3.8), such that $\mathrm{Re}\lambda_1 \geq \mathrm{Re}\lambda_2$.

The corresponding Fuchsian system (at the moment "Fuchsianity" means Fuchsianity at 0) is a system for an unknown vector

$$\vec{y} = \begin{pmatrix} y \\ x\frac{dy}{dx} \end{pmatrix}.$$

This system has the following form:

$$\frac{d\vec{y}}{dx} = \frac{1}{x}C(x)\vec{y}, \quad C(x) = \begin{pmatrix} 0 & 1 \\ -q(x) & 1 - p(x) \end{pmatrix}. \tag{7.3.9}$$

A characteristic equation for $C(0)$ is just (7.3.8).

Let $Y(\tilde{x}) = (\vec{y}_1(\tilde{x}), \vec{y}_2(\tilde{x}))$ be a Levelt's fundamental system of solutions and let

$$Y(\tilde{x}) = V(x)x^\Phi \tilde{x}^E = (\vec{v}_1(x), \vec{v}_2(x))x^{\begin{pmatrix} \varphi^1 & 0 \\ 0 & \varphi^2 \end{pmatrix}}\tilde{x}^{\begin{pmatrix} \mu^1 & e \\ 0 & \mu^2 \end{pmatrix}} \tag{7.3.10}$$

be its factorization of form (7.2.1). (Recall that E is upper triangular, e can be 0 in one cases and $\neq 0$ in other. We know that matrices $C(0), L(0), \Phi + E$ are similar. Thus, $\lambda_j = \varphi^j + \mu^j$ which provides the better motivation for (7.3.8) from the point of view of our theory than the arguments presented above: (7.3.8) allows to determine φ^j and μ^j. Namely, let $\lambda_j = \beta_j + i\gamma_j$. As $0 \leq \mathrm{Re}\mu^j < 1$, we get $\varphi^j = [\beta_j]$, $\mu^j = \{\beta_j\} + i\gamma_j$, where $[x]$, and $\{x\}$ denote the entire and nonentire parts of the number x respectively. (Cf. 2.2.2). It follows from (7.3.10) that

$$\vec{y}_1(\tilde{x}) = x^{\varphi^1}\tilde{x}^{\mu^1}\vec{v}_1(x),$$

while the corresponding expression for $\vec{y}_2$ may contain a logarithm (if $e \neq 0$). Taking into account what we know about Levelt's valuations, we obtain the following conclusion:

i) equation (7.3.7) always has a solution of the form $\tilde{x}^{\lambda_1}f(x)$, where f is holomorphic in some disk with the center at 0, and $f(0) \neq 0$. (If there were $f(0) = 0$, then it would follow that $\varphi(\tilde{x}^{\lambda_1}f(x)) > \varphi^1$).

In addition to it we shall prove the following statement:

ii) if $\lambda_1 - \lambda_2$ is not integer, then (7.3.7) has also a solution of the form $\tilde{x}^{\lambda_2} g(x)$, where $g(x)$ has the same properties as f.

Consider the first case : $\varphi^1 = \varphi^2$. Then Levelt's filtration reduces to the following trivial one:

$$\{0\} = X^0 \subset X^1 = X,$$

i.e. all nontrivial solutions have the same Levelt's valuation. So all we need to obtain a Levelt's basis is to choose $\vec{y}_1, \vec{y}_2$ such that the monodromy matrix G (which describes the map σ^* in terms of the basis $\vec{y}_1, \vec{y}_2$) would be upper-triangular. But in the case under consideration G can be transformed into diagonal form. Indeed, since

$$\mu^1 - \mu^2 = \varphi^1 - \varphi^2 + \mu^1 - \mu^2 = \lambda_1 - \lambda_2,$$

we have that $\mu^1 - \mu^2$ is not integer. Thus, eigenvalues $e^{2\pi i \mu^j}$, $j = 1, 2$ of the matrix G are different. This implies that $e = 0$ in (7.3.10) and $\vec{y}_2(\tilde{x}) = = x^{\varphi^2} \tilde{x}^{\mu^2} \vec{v}_2(x)$, consequently (7.3.7) has the second solution of the desired form and this solution is linearly independent of the first one.

Consider now the case $\varphi^1 > \varphi^2$. Levelt's filtration has the form

$$\{0\} = X^0 \subset X^1 \subset X^2 = X, \quad \dim X^1 = 1.$$

A vector $\vec{y}_1$ belonging to X^1 is defined in essentially unique way (up to a multiplier) and there is a freedom in the choice of $\vec{y}_2$ from $X^2 \setminus X^1$. For any choice of such $\vec{y}_2$ we obtain an upper-triangular G and E. Indeed, since $\lambda_1 - \lambda_2 = \varphi^1 - \varphi^2 + \mu^1 - \mu^2$ is noninteger and φ^j are integer, it follows that $\mu^1 - \mu^2$ is noninteger number, so that eigenvalues of G are different and G has two linearly independent eigenvectors. We know that $\vec{y}_1$ is one of them, thus we can take $\vec{y}_2$ to be another eigenvector.

We obtained i) and ii) as immediate consequences of Levelt's theory, but classically they were proved in another way (and long before Levelt). Classical proofs can be found in many textbooks, e.g. [Ha], [CoLe], [Tr]. Tricomi considers only the case of 2nd order Fuchsian ODE (which we need in this section), whereas other two books contain a more complicated treatment of Fuchsian systems of any order.

3. Now we turn to Bessel's equation

$$\frac{d^2 y}{dx^2} + \frac{1}{x} \frac{dy}{dx} + (1 - \frac{\nu^2}{x^2}) y = 0 \qquad (7.3.11)$$

(ν is a parameter). Its solutions are called *Bessel's functions* (so to say, in a broad sense). There are special solutions which have special names. One of them is also called Bessel's function (so to say, in the narrow sense).

Equation (7.3.11) has a regular singular point $x = 0$ and irregular singular point $x = \infty$. The indical equation at $x = 0$ is as follows:

$$\lambda^2 - \nu^2 = 0. \tag{7.3.12}$$

According to general theory , one can consider solutions to (7.3.11) as functions on $\tilde{S}$, which is an universal covering surface for $S = \bar{\mathbb{C}} \setminus \{0, \infty\} = \mathbb{C} \setminus \{0\}$. Thus, it would be more correct to write $y(\tilde{x})$, as we do in numerous cases. However, in this section we make a concession to classical treatments and write $y(x)$, when dealing with the Bessel functions.

We shall study a behavior of solutions to (7.3.11) near $x = 0$. Thus, our approach is local. But since there are no other singularities in $\mathbb{C}$, our results will be not so local. Formulas we shall write will be valid in the whole $\mathbb{C}$
(to be precise, in $\tilde{S}$) and the functions appearing there will be either entire functions or meromorphic functions having only pole at $x = 0$ up to multipliers like x^ν and up to change of independent variable like something with the square root $x = 2\sqrt{t}$
. (Of course, this is the simplest part of the theory of Bessel's functions. Another part concerns their behavior at ∞. Also, there are approaches to this theory based not on equation (7.3.11), but on the integral representations for these functions and on the theory of group representations).

Formula (7.3.12) prompts the change $y = x^\nu z$ of an unknown function. One gets the following equation for z:

$$\frac{d^2 z}{dx^2} + \frac{1 + 2\nu}{x} \frac{dz}{dx} + z = 0, \tag{7.3.13}$$

and the corresponding indical equation at the point 0:

$$\lambda(\lambda + 2\nu) = 0. \tag{7.3.14}$$

Applying i) and ii), we obtain the following three cases:

1) if 2ν is not integer, then (7.3.13) has solutions of the next forms: $f(x)$ and $x^{2\nu} g(x)$;

2) if $2\nu \in \mathbb{Z}$ and $\nu \geq 0$, then (7.3.13) has a solution of the form $f(x)$;

3) if $2\nu \in \mathbb{Z}$ and $\nu < 0$, then (7.3.13) has a solution of the form $x^{2|\nu|} f(x)$.

Above, f and g are entire, because 0 is the unique singular point of the differential equation (7.3.13) in $\mathbb{C}$.

Now pay attention to the parity properties of the coefficients of (7.3.13) : a coefficient at $\frac{dz}{dx}$ is odd and other coefficients are even. First of all we shall try to use this fact in the most "naive", "immediate" way; later we shall use it more effectively.

Immediate conclusions concern the existence of even entire solutions. Parity of coefficients implies that if $z(x)$ is a solution to (7.3.13), then $y(x) := z(-x)$ is also a solution. Thus,

$$f(x) + f(-x) \tag{7.3.15}$$

in the first two cases and

$$x^{2|\nu|} f(x) + (-x)^{2|\nu|} f(-x) \tag{7.3.16}$$

in the third case also are solutions. They are entire and even. Solution (7.3.15) is nontrivial, because $f(0) \neq 0$. Solution (7.3.16) could be identically equal to zero only if $x^{2|\nu|} f(x)$ were odd. Since $f(0) \neq 0$, this would imply that $2|\nu|$ were odd. Thus, in the case, when 2ν is a negative integer odd number it remains uncertain whether (7.3.13) has an entire even solution.

Later we shall see that a more clever use of the parity of the coefficients of (7.3.13) provides a positive answer to the last question. And in fact, we shall see a more important thing: in the case , when 2ν is a negative integer odd number (independently of the sign of ν), (7.3.13) has solutions of the same form as in the case 1) (so one of them again is of the form $f(x)$ with entire even function f, $f(0) \neq 0$), whereas in the case $\nu \in \mathbb{Z}$ equation (7.3.13) has a logarithmically branching solution (and as we already know, it also has an entire even solution of the form $f(x)$ for $\nu = 0$ and $x^{2|\nu|} f(x)$ for $\nu < 0$). The arguments below from the very beginning assure that f, g are even (so the primitive arguments above with (7.3.15), (7.3.16) are needless in reality).

4. Return to equation (7.3.7) and assume that p and q are even functions (of course, holomorphic in some neighborhood of 0). Introduce a new independent variable t such that $x = 2\sqrt{t}$. (The multiplier 2 is not essential, simply it is convenient for calculations). Dealing with (7.3.7), we shall return to "less classical" notation, where S is replaced by $\tilde{S}$, x is replaced by $\tilde{x}$ etc. When we write $x = 2\sqrt{t}$, this means that we have a two-sheeted covering $S \to \Sigma = \mathbb{C} \setminus \{0\}$, $x \in S$, $t \in \Sigma$. Of course, this Σ is the same Riemann surface as S, but when we consider a covering $S \to S$, it seems to be more convenient in our case to distinguish these two copies of S. The only nontrivial deck transformation for $S \to \Sigma$ is the following one : $i : x \mapsto -x$. An even function on S is the function such that it can be obtained as a lifting of some function on Σ. By virtue of the composition $\tilde{S} \to S \to \Sigma$, $\tilde{S}$ becomes the universal covering of Σ; when considered in such a role, $\tilde{S}$ will be denoted by $\tilde{\Sigma}$. In terms of t we obtain the equation

$$\ddot{y} + \frac{1 + p(2\sqrt{t})}{2t} \dot{y} + \frac{q(2\sqrt{t})}{4t^2} y = 0, \tag{7.3.17}$$

where a derivation with respect to t is denoted by dot. Since p and q are even, the numerators in (7.3.17) are holomorphic functions in t in some neighborhood of 0. Thus, (7.3.17) is of the same type as (7.3.7). When dealing with $y(x(t))$, we again write $y(t)$, which can be considered as a kind of quite common freedom of speech. But if we want to be precise, we must remember that solutions to (7.3.7), (7.3.17) are not single-valued functions in x, t; they are functions on $\tilde{S} = \tilde{\Sigma}$ (on some part of $\tilde{S} = \tilde{\Sigma}$, which is a covering of some punctured neighborhood of 0 in Σ). And even from the most formal point of view there is no harm in writing $y(\tilde{t})$ instead of $y(\tilde{x})$,–as $\tilde{S} = \tilde{\Sigma}$, $\tilde{x}$ and $\tilde{t}$ may well be the different notation to the same point (which covers points $x \in S$ and $t = \frac{x^2}{4} \in \Sigma$). (So some freedom of speech is not in writing $y(t)$ instead of $y(x(t))$, but in writing $y(x)$ instead of $y(\tilde{x})$!).

If λ_1, λ_2 are solutions to the indical equation for the original equation (7.3.7), then the solutions to the indical equation for (7.3.17) are $\lambda_1/2$, $\lambda_2/2$. This is clear even without wring out the latter indical equation, just in view of the formal way of obtaining the indical equation, because the former x^λ (in reality, $\tilde{x}^\lambda$) now becomes $t^{\lambda/2}$.

Now, if $\lambda_1/2 - \lambda_2/2$ is noninteger, then i) assures that (7.3.17) has two linearly independent solutions of the form

$$\tilde{t}^{\lambda_1/2} f_1(t), \quad \tilde{t}^{\lambda_2/2} g_1(t)$$

with f_1, g_1 holomorphic at 0 and $f_1(0) \neq 0$, $g_1(0) \neq 0$. Thus, the parity properties of p, q allow somewhat to strengthen i) and ii):

iii) if $\lambda_1 - \lambda_2$ is not integer, then (7.3.7) has solutions of the form $\tilde{x}^{\lambda_1} f(x)$, $\tilde{x}^{\lambda_2} g(x)$, where f, g are even and $f(0) \neq 0$, $g(0) \neq 0$.

Of course, these f, g are entire if, in addition to our assumptions, p and q are entire functions.

5. Applying these considerations to (7.3.13), we come to the equation

$$\ddot{z} + \frac{1 + \nu}{t} \dot{z} + \frac{1}{t} z = 0 \qquad (7.3.18)$$

with the indical equation $\lambda(\lambda + \nu) = 0$. If ν is noninteger, then (7.3.18) has an entire solution, corresponding to $\lambda = 0$ and also a solution of the form $t^{-\nu} \cdot$ (entire function), corresponding to $\lambda = -\nu$. If $\nu \in \mathbb{Z}$ and $\nu \geq 0$, then (7.3.18) has an entire solution, corresponding to $\lambda = 0$. If $\nu \in \mathbb{Z}$ and $\nu < 0$, then (7.3.18) has a solution of the form

$$t^{-\nu} \cdot \text{(entire function)}, \qquad (7.3.19)$$

which is also an entire function. However, the previous entire functions do not vanish at $t = 0$; and (7.3.19) equals zero at $t = 0$.

Let $z = \sum_{n=0}^{\infty} c_n t^n$ be an entire solution. Substituting this into (7.3.18), one obtains the recurrent relations

$$n(n + \nu)c_n + c_{n-1} = 0, \quad n = 1, 2, \ldots \qquad (7.3.20)$$

If $\nu = -l$, $l \in \mathbb{Z}$, $l > 0$, then it follows that $n + \nu = 0$ for $n = l$, so $c_{l-1} = 0$. Then, (7.3.20) implies that $c_n = 0$ for all $n < l$. In this case we can start from arbitrary c_l and then define c_n for $n > l$, using (7.3.20). For other ν we can start from arbitrary c_0 and then define c_n for all $n > l$. A particular choice of c_l or c_0 means a kind of "normalization". "Standard" ways of "normalization" in the theory of differential equations are related either to the Cauchy matrix, which is the identity matrix at some prescribed point or to orthogonality. In these cases one performs "normalization" having in mind some exact criterion. But for the special functions "normalization" often means a more simple form of their coefficients or some formulas,–a criterion , which scarcely can be formalized. In the theory of Bessel's functions the following "normalization" is used:

$$\text{if } \nu \in \mathbb{Z}, \ \nu < 0, \ \nu = -l, \text{ then } c_l = \frac{(-1)^l}{l!};$$

$$\text{otherwise } c_0 = \frac{1}{\Gamma(\nu + 1)}, \text{ where } \Gamma \text{ is Euler's gamma-function.}$$

Then one obtains the following series:

$$F_\nu(t) := \sum_{n=0}^{\infty} \frac{(-1)^n}{\Gamma(n + 1)\Gamma(n + \nu + 1)} t^n. \qquad (7.3.21)$$

(If $\nu = -l$, then the first l coefficients have infinite denominator, so this series starts from $\frac{(-1)^l}{l!} t^l$. It is remarkable that in this "exceptional" case one manages to write the same formula as in the "typical" case).

We shall use later that the map

$$\mathbb{C} \times \mathbb{C} \to \mathbb{C}, \ (\nu, t) \mapsto F_\nu(t) \qquad (7.3.22)$$

is holomorphic. This is proved as follows. It is well known that $1/\Gamma(s)$ is an entire function in s. Thus,

$$f_k(t, \nu) := \sum_{n=0}^{k} \frac{(-1)^n t^n}{\Gamma(n + 1)\Gamma(n + 1 + \nu)}$$

is an entire function in (t, ν). It is sufficient to prove that for any $r > 0$ the convergence $f_k(t, \nu) \to F_\nu(t)$ is uniform with respect to $|t|$, $|\nu| \leq r$. Fix $j > r$. Let M be such that

$$\left| \frac{1}{\Gamma(j+1+\nu)} \right| < M \text{ for all } \nu \text{ with } |\nu| \le r.$$

Then for $|\nu| \le r$ and integer $n > j$ we have

$$|n + \nu| \ge \operatorname{Re}(n + \nu) \ge n - r \ge n - j \ge 1,$$

$$\left| \frac{1}{\Gamma(n+1+\nu)} \right| = \frac{1}{|(n+\nu)\cdots(j+1+\nu)\Gamma(j+1+\nu)|} \le M.$$

Thus, if $m > l > 0$ and $|t|, |\nu| \le r$, then

$$|f_{j+m}(t,\nu) - f_{j+l}(t,\nu)| =$$

$$\left| \sum_{n=j+l+1}^{j+m} \frac{(-1)^n t^n}{\Gamma(n+1)\Gamma(n+1+\nu)} \right| \le M \sum_{n=j+l+1}^{j+m} \frac{r^n}{n!},$$

which proves the uniform convergence.

Returning to (7.3.11), one gets a special solution to (7.3.11), which is called *Bessel's function* (in the narrow sense), it is also called *Bessel's functions of the first kind* and is defined as follows:

$$J_\nu(x) = t^{\nu/2} F_\nu(t) = (x/2)^\nu F_\nu(x^2/4). \tag{7.3.23}$$

Usually one deals directly with this function. However, from the local point of view we explain here F_ν seems to be somewhat more convenient: it never branches, series (7.3.21) is slightly simpler than the corresponding expression for J_ν. (But we do not claim that F_ν is more convenient for all purposes).

Notation F_ν is used in the book [GM]. These authors attribute this function to G.Greenhill. Tricomi in [Tr] also uses this function, but he denotes it by $E_\nu(t)$.

In (7.3.11) we can replace ν by $-\nu$ without any change. We never used any condition on ν like fixing its sign or sign of $\operatorname{Re}\nu$. Thus, $J_{-\nu}(x)$ is a solution to (7.3.11) as well as $J_\nu(x)$. If ν is not integer, then

$$J_{-\nu}(x) = x^{-\nu} \left(\frac{2^\nu}{\Gamma(-\nu+1)} + \dots \right),$$

$$J_\nu(x) = x^\nu \left(\frac{1}{2^\nu \Gamma(\nu+1)} + \dots \right),$$

where dots denote the convergent power series in x (in fact, in x^2) without free term. Clearly, $J_\nu(x)$ and $J_{-\nu}(x)$ are linearly independent; so we found two linearly independent solutions to (7.3.11). In different terms we can say that

$$\left(\frac{x}{2}\right)^{-\nu} J_\nu(x), \ \left(\frac{x}{2}\right)^{-\nu} J_{-\nu}(x)$$

are solutions to (7.3.13), and

$$t^{-\nu/2} J_\nu(2\sqrt{t}) = F_\nu(t), \ \ t^{-\nu/2} J_{-\nu}(2\sqrt{t}) = t^{-\nu} F_\nu(t)$$

are solutions to (7.3.18). If ν is noninteger, then

$$F_\nu(t) = \frac{1}{\Gamma(\nu+1)} + \dots, \ \ t^{-\nu} F_{-\nu}(t) = \frac{t^{-\nu}}{\Gamma(-\nu+1)} + \dots$$

are linearly independent solutions to (7.3.18).

Let us summarize what we proved for noninteger ν. We may assume that $\mathrm{Re}\,\nu \geq 0$, because (7.3.7) contains only ν^2. Then $(J_\nu, J_{-\nu})$ is Levelt's fundamental system of solutions to (7.3.11). The related matrices E and Φ are diagonal and if $\nu = \beta + i\gamma$, $\beta, \gamma \in \mathbb{R}$, then $\varphi^1 = [\beta]$, $\mu^1 = \{\beta\} + i\gamma$, $\varphi^2 = -[\beta] - 1$ for noninteger β and $\varphi^2 = -[\beta]$ for $\beta \in \mathbb{Z}$ (in the present case this happens only when $\gamma \neq 0$), $\mu^2 = 1 - \{\beta\} - i\gamma$ for noninteger β, and $\mu^2 = -i\gamma$ for $\beta \in \mathbb{Z}$.

The essential new information we got comparatively to item 3 is that the case $\nu = k + 1/2$, $k \in \mathbb{Z}$ is not exceptional as far as concerns the existence of solutions of the special type we discussed. By the way, it is exceptional in another sense: in this case Bessel's function can be expressed by means of elementary ones.

If $\nu = -1/2$, (7.3.13) becomes $\frac{d^2 z}{dx^2} + z = 0$. This equation has a fundamental system of solutions $\cos x, \sin x$. It turns out that

$$F_{-\frac{1}{2}}(t) = F_{-\frac{1}{2}}\left(\frac{x^2}{4}\right), \ \ t^{\frac{1}{2}} F_{\frac{1}{2}}(t) = \frac{x}{2} F_{\frac{1}{2}}\left(\frac{x^2}{4}\right),$$

up to a multiplier are just these functions: some computation with help of (7.3.21) and well known properties of Γ shows that

$$F_{-\frac{1}{2}}\left(\frac{x^2}{4}\right) = \frac{1}{\sqrt{\pi}} \cos x, \ \ F_{\frac{1}{2}}\left(\frac{x^2}{4}\right) = \frac{2}{\sqrt{\pi x}} \sin x.$$

And (7.3.23) gives elementary expression for $J_{\mp 1/2}$. In order to see that other $J_{k+1/2}$ also have elementary expressions, one can use the recurrent relation

$$J_{\nu+1} + J_{\nu-1} = \frac{2\nu}{x} J_\nu, \tag{7.3.24}$$

which can be proved as follows. First, a calculation using (7.3.21), shows that $\dot{F}_\nu(t) = -F_{\nu+1}(t)$. Thus, it follows from (7.3.18) that

$$F_{\nu+2} - \frac{1+\nu}{t}F_{\nu+1} + \frac{1}{t}F_{\nu} = 0.$$

After we replace ν by $\nu - 1$, the latter relation becomes equivalent to (7.3.24).

6. Let now $\nu \in \mathbb{Z}$. In this case we usually write n instead of ν. We can assume that $n \geq 0$. We have

$$F_{-n}(t) = \sum_{j=n}^{\infty} \frac{(-1)^j t^j}{\Gamma(j+1)\Gamma(j-n+1)} =$$

$$= \sum_{m=0}^{\infty} \frac{(-1)^{n+m} t^{n+m}}{\Gamma(n+m+1)\Gamma(m+1)} = (-1)^n t^n F_n(t).$$

After we proved this for $n \geq 0$, it is easy to see that the formula

$$F_{-n}(t) = (-1)^n t^n F_n(t) \tag{7.3.25}$$

is valid for all $n \in \mathbb{Z}$. Thus, $t^{-n} F_{-n}(t) = (-1)^n F_n(t)$ is the same solution to (7.3.18) as $F_n(t)$ up to sign. In terms of J_n (7.3.25) means

$$J_{-n}(x) = (-1)^n J_n(x). \tag{7.3.26}$$

Later we shall construct a solution to (7.3.18) (with $\nu = n$) which branches logarithmically: it would follow that the only solutions which do not branch logarithmically are multiples of F_n. But it is instructive to give a direct proof of the latter fact using only "general consideration".

Roots of the indical equation for (7.3.18) are $\lambda_1 = 0$, $\lambda_2 = -n$, so $\varphi^1 = 0$, $\varphi^2 = -n$, $\mu^j = 0$. Writing a Fuchsian system (7.3.9) for (7.3.18), we find that

$$C(0) = \begin{pmatrix} 0 & 1 \\ 0 & -n \end{pmatrix}.$$

Recall that in (7.3.10) $\Phi + E$ is similar to $C(0)$.

If $n = 0$, then $C(0)$ is the Jordan block of the second order, $\Phi = 0$ and therefore E is similar to such a block. Thus, e can not be equal to 0, and the second solution from the Levelt's fundamental system branches logarithmically.

If $n > 0$, then the argument must be somewhat more subtle , because $C(0)$ is diagonalizable and the fact that

$$\begin{pmatrix} 0 & 0 \\ 0 & n \end{pmatrix} + E$$

is similar to it does not provide enough information. Assume that (7.3.18) has a single-valued solution z, which is linearly independent of F_n. Its valuation must be $\varphi(z) = -n$. Thus, $z = \sum_{j=-n}^{\infty} c_j t^j$, $c_{-n} \neq 0$,i.e. $z = t^{-n}w$ with entire w. Then w must be a solution to (7.3.18) with n replaced by $-n$, i.e. solution to

$$\ddot{w} + \frac{1-n}{t}\dot{w} + \frac{1}{t}w = 0. \tag{7.3.27}$$

This can be easily checked immediately, but in reality we already know this. Indeed, if z is the solution to (7.3.18) (with ν replacing by n), then $z(x^2/4) = x^{-n}y(x)$, where y is a solution to (7.3.11). Thus, $y = \text{const} \cdot x^{-n}w$; const is of no importance. We see that the passage from y to w is the same change of dependent variable as the change $y = x^n z$, which we considered in item 3, only with $-n$ instead of n. (Remember that in (7.3.11) one can replace ν by $-\nu$). Thus, w satisfies (7.3.13) with n being replaced by $-n$, i.e.

$$\frac{d^2z}{dx^2} + \frac{1-2n}{x}\frac{dz}{dx} + z = 0, \tag{7.3.28}$$

and then, as we know, in terms of the independent variable t, w is a solution to (7.3.27), the latter being (7.3.18) with ν being replaced by $-n$.

In item 5 we tried to find a power series which is a solution to (7.3.18) and found out that in the case $\nu = -l$, $l \in \mathbb{Z}$, $l > 0$ the only solution is $\text{const} \cdot F_{-l}$. For (7.3.27) this means that the only solution to it, which is holomorphic in some neighborhood of 0 is a multiple to F_n; its first term is $c_n t^n$ and not the c_0. In terms of (7.3.28) this means that this equation has no solutions of the form

$$z = x^{-2n}(c_0 + c_2 x^2 + c_4 x^4 + \cdots), \ c_0 \neq 0,$$

where the term in the brackets is an even function which is holomorphic in some neighborhood of 0. We want to prove a somewhat more general statement: (7.3.28) has no solution of the form

$$z = x^{-2n}(c_0 + c_1 x + c_2 x^2 + c_3 x^3 + \cdots), \ c_0 \neq 0.$$

Assume that there exists such a solution. Then $z(-x)$ is also the solution (because in (7.3.28) the coefficient at z is odd and other coefficients are even), and so the even function

$$z(x) + z(-x) = x^{-2n}(2c_0 + 2c_2 x^2 + 2c_4 x^4 + \cdots)$$

would also be a solution, which is contradictory.

If in (7.3.10) there were $e = 0$, then E would be equal to 0, because we know that $\mu^j = 0$.

Thus, (7.3.13) would have a Levelt's fundamental system of solutions, which would be single-valued on S. (Conversely, if there were two linearly independent single-valued solutions on S, then σ^* would be an identity map of X, thus in any basis there would be $E = 0$). In Levelt's basis the second solution would be of the form $x^{-2n}(c_0 + c_1 x + \cdots)$, $c_0 \neq 0$. But (7.3.13) does not have such a solution. So,

$$E = \begin{pmatrix} 0 & e \\ 0 & 0 \end{pmatrix} \text{ with } e \neq 0.$$

Returning to the beginning of item 2, we obtain in the corresponding(7.3.10)

$$(\vec{y}_1, \vec{y}_2) = (\vec{v}_1, \vec{v}_2) x^{\begin{pmatrix} n & 0 \\ 0 & -n \end{pmatrix}} \tilde{x}^{\begin{pmatrix} 0 & e \\ 0 & 0 \end{pmatrix}} =$$

$$= V(x) \begin{pmatrix} x^n & 0 \\ 0 & x^{-n} \end{pmatrix} \begin{pmatrix} 1 & e \ln \tilde{x} \\ 0 & 1 \end{pmatrix}, \ G = \begin{pmatrix} 1 & 2\pi e i \\ 0 & 1 \end{pmatrix}.$$

By changing $\vec{y}_2$, one can make e any given number except 0. Prescribing this number defines $\vec{y}_2$ up to a summand from $\mathbb{C}\vec{y}_1$. A standard choice of $\vec{y}_2$ is provided by a special kind of the second solution to (7.3.11) – the so-called *Neumann-Veber function* (or *Bessel's function of the second kind*). It is usually denoted by N_ν or Y_ν; we shall use the first notation.

N_ν is defined for all ν, not only integers. First, we shall give its definition for noninteger ν:

$$N_\nu := \frac{1}{\sin \pi \nu} (J_\nu(x) \cos \pi \nu - J_{-\nu}(x)). \tag{7.3.29}$$

Clearly, it is solution to (7.3.11). Replace ν by $-\nu$ in (7.3.27) and consider the two relations thus obtained as a system of 2 linear algebraic equations which allows "unknowns" $J_{-\nu}, N_{-\nu}$ in terms of J_ν, N_ν. As a result we get

$$J_{-\nu} = J_\nu \cos \pi \nu - N_\nu \sin \pi \nu, \ N_{-\nu} = J_\nu \sin \pi \nu + N_\nu \cos \pi \nu. \tag{7.3.30}$$

This is easy to remember, because it looks like a rotation in X. However, it is not a theorem, but rather a matter of definition –having two independent vectors J_ν and $J_{-\nu}$, we define new vectors $N_\nu, N_{-\nu}$ in such a way that (7.3.30) holds.

It turns out that for any integer n and $x \neq 0$ there exists $\lim_{\nu \to n} N_\nu(x)$, which is taken as definition for $N_n(x)$. In virtue of analyticity of J_ν with respect to ν (when $x \neq 0$) we can use L'Hopitale's rule, which gives

$$N_n = \frac{1}{\pi} \left[\frac{\partial}{\partial \nu} J_\nu - (-1)^n \frac{\partial}{\partial \nu} J_{-\nu} \right]_{\nu=n}. \tag{7.3.31}$$

Convergence in $N_\nu \to N_n$ is uniform with respect to any compact $K \subset S$. Thus, N_ν is a solution to (7.3.11) (with $\nu = n$).

Similarly to (7.3.26),

$$N_{-n}(x) = (-1)^n N_n(x).$$

This follows directly from (7.3.31)–one does not even use (7.3.26). (Whereas the proof of (7.3.26) is more "susbstantional"–one uses a precise form of the power series for F_ν). So for an integer ν (when there is a special need in N_ν) it is sufficient to consider only $N_{|\nu|}$.

Now our goal is to present N_n, $n \in \mathbb{Z}$, $n \geq 0$, in a form $A(\ln x) J_n(x) +$ meromorphic function and to compute this A. In particular, we shall show that $A \neq 0$. Thus, N_n is really solution to (7.3.11), which is linearly independent of J_n.

Apparently it is again better to work in terms of (7.3.18) and its solutions. Since each solution to (7.3.11) is equal to some solution to (7.3.18), multiplied by $t^{-\nu/2}$ (recall that $x = 2\sqrt{t}$), we shall define $\Xi_\nu(t)$ in such a way that

$$N_\nu(x) = (x/2)^\nu \Xi_\nu(x^2/4) = t^{\nu/2} \Xi_\nu(t),$$

i.e. for noninteger ν we have

$$\Xi_\nu(t) = \frac{1}{\sin \pi\nu}(F_\nu(t) \cos \pi\nu - t^{-\nu} F_{-\nu}(t)).$$

(Notation Ξ_ν is by no means common. We introduce it only for a minute. We intentionally use a letter which is "free" in this area so that our use of it can not lead to misanderstanding). It's clear that

$$\Xi_n(t) = \frac{(-1)^n}{\pi}\left[(-1)^n\frac{\partial F_\nu}{\partial \nu} - t^{-n}\frac{\partial F_{-\nu}}{\partial \nu} - F_{-\nu}\frac{\partial t^{-\nu}}{\partial \nu}\right]_{\nu=n} =$$

$$= \frac{1}{\pi}\left[\frac{\partial F_\nu(t)}{\partial \nu} - (-1)^n t^{-n}\frac{\partial F_{-\nu}(t)}{\partial \nu}\right]_{\nu=n} - \frac{1}{\pi}\underline{(-1)^n t^{-n} F_{-n}(t)}(-\ln t).$$

By virtue of (7.3.25), the underlined term equals $F_n(t)$. Remember that (7.3.22) is holomorphic. Thus, the square bracket is equal to some $\sum_{j=-n}^{\infty} a_j t^j$. Now $\ln t = 2\ln x - 2\ln 2$, so

$$N_n(x) = (x/2)^n \Xi_n(t) = \sum_{j=-n}^{\infty} b_j x^j + \frac{2}{\pi}(\ln x)J_n(x). \tag{7.3.32}$$

Usually one writes this in a slightly different form:

$$N_n(x) = \frac{2}{\pi}(\ln(x/2) + C)J_n(x) + \sum_{j=-n}^{\infty} c_j x^j, \qquad (7.3.33)$$

where C is the Euler's constant. When passing from (7.3.32) to (7.3.33), one adds $\frac{2}{\pi}(C - \ln 2)J_n(x)$ to $\frac{2}{\pi}\ln x J_n(x)$ and subtracts the same term from $\sum_{j=-n}^{\infty} b_j x^j$. Explicit formulas for c_j are known, but they are more complicated than in (7.3.21) and we shall not write them.

Logarithmic term makes clear that N_n is linearly independent of J_n. Then, as we know, there must be $\varphi(N_n) = -n$. Thus, for $n > 0$ b_{-n} in (7.3.32) and c_{-n} in (7.3.33) are not equal to zero. The monodromy is clear from (7.3.33):

$$N_n(\sigma\tilde{x}) = \frac{2}{\pi}(\ln(\sigma\tilde{x}) - \ln 2 + C)J_n(x) + \sum_{j=-n}^{\infty} b_j x^j = N_n(\tilde{x}) + 4iJ_n(x),$$

$$G = \begin{pmatrix} 1 & 4i \\ 0 & 1 \end{pmatrix}, \quad E = \begin{pmatrix} 0 & \frac{2}{\pi} \\ 0 & 0 \end{pmatrix}.$$

There is a generally accepted agreement that Bessel'a functions of the first kind are just the J_ν described above. As regards to Bessel's functions of the second kind, at least 2 linear combinations of our N_ν and J_ν are used as well (besides, there is a version of N_ν, which differs from our N_ν by a multiplier depending on ν). Primarily this is a matter of hystorical accidence, but also up to some extent this seems to be related to the fact that usually the first vector of Levelt's basis is uniquely defined up to a multiplier, whereas there is more freedom in the choice of the second vector. (Of course, there can be also reasons of the different nature,–e.g., functions $J_\nu \mp iN_\nu$ have simpler asymptotic at infinity).

Bibliography

[An] Anosov, D.V., An introduction to Hilbert's 21st problem, preprint of the Institute for Mathematics and its Applications, University of Minnesota, IMA Preprint Series **861** (1991).

[ArIl] Arnold, V.I. and Il'yashenko, Yu.S., Ordinary differential equations, in the book: Anosov, D.V. and Arnold, V.I. (eds.), Dynamical systems 1, Encyclopaedia of the Mathematical sciences, 1, Springer, 1988.

[At] Atiyah, M.F., Complex analytic connections in fibre bundles, Trans. Amer. Math. Soc., **85** (1957), 181–207.

[Bi] Birkhoff, G.D., Collected mathematical papers , 1,Dover Publ., Inc., New-York, 1968.

[Boj] Bojarski, B., Connections between Complex and Global Analysis, Complex Analysis. Methods. Applications, A.V., Berlin, 1983.

[Bo1] Bolibruch, A.A., The Riemann-Hilbert problem on the complex projective line (in russian), Mat. Zametki **46:3** (1989), 118–120.

[Bo2] Bolibruch, A.A., The Riemann-Hilbert problem, Russian Math. Surveys, **45:2** (1990), 1–47.

[Bo3] Bolibruch, A.A., Construction of a Fuchsian equation from a monodromy representation, Mathem. Notes of Ac. of Sci. of USSR **48:5** (1990), 1090–1099.

[Bo4] Bolibruch, A.A., Fuchsian systems with reducible monodromy and the Riemann-Hilbert Problem, Lecture Notes in Mathem., **1520**, Springer, 1992, 139–155.

[Bo5] Bolibruch, A.A., On sufficient conditions for the positive solvability of the Riemann-Hilbert Problem, Mathem. Notes of Ac. of Sci. of USSR , **51:2** (1992), 110–117.

[Bo6] Bolibruch, A.A., Hilbert's twenty-first problem for Fuchsian linear systems, in: Arnold, V. and Monastyrsky, M., (eds.) Developments in Mathematics: The Moscow School, Chapman and Hall, London – Madras, 1993, 54–99.

[CoLe] Coddington, E.A. and Levinson, N., Theory of ordinary differential equations, McGraw-Hill, New York–London, 1955.

[Dek] Dekkers, W., The matrix of a connection having regular singularities on a vector bundle of rank 2 on $P^1(\mathbb{C})$, Lecture Notes in Mathem., **712**, Springer, 1979, 33–43.

[Del] Deligne, P., Equations differentielles a points singuliers reguliers, Lecture Notes in Mathem., **163**, Springer, 1970.

[EsVi] Esnault, H. and Viehweg, E., Logarithmic De Rham comlexes and vanishing theorems, Inv. Math., **86** (1986), 161–194.

[Fö] Förster, O., Riemannische Flächen, Springer, Berlin, 1970.

[Ga] Gantmacher, F.R., Theory of matrices, II, Chelsea, New York, 1959.

[Ge] Gerard, R., Le problème de Riemann-Hilbert sur une variété analytic complex, Ann. Inst. Fourier, **19** (1969), 1–12.

[GM] Gray, A. and Mathews, F., A treatise on Bessel functions and their application to physics, 1931.

[Go1] Golubeva, V.A., Reconstruction of a Pfaff system of Fuchs type from the generators of the monodromy group, Math. USSR-Izv., **17** (1981), 227–241.

[Go2] Golubeva, V.A., Some problems in the analytic theory of Feynman integrals, Russian Math. Surveys, **31:2** (1976), 135–202.

[Ha] Hartman, P., Ordinary differential equations, John Wiley, 1964.

[Hai] Hain, R.M., On a generalization of Hilbert's 21st problem, Ann. Sci. Ecole Norm. Sup., **19:4** (1986), 609–627.

[Hi1] Hilbert, D., Mathematische Probleme, Nachr. Ges. Wiss., Göttingen, 1900, 253–297.

[Hi2] Hilbert, D., Grundzüge einer allgemeinen Theorie der linearen Integralgleichungen (Dritte Mitt.), Nachr. Ges. Wiss., Göttingen, 1905, 307–338.

[In] Ince, E.L., Ordinary differential equations, Dover, 1956.

[Ka] Katz, N.M., An overview of Deligne's work on Hilbert's twenty-first problem, Proc. Sympos. Pure Math., **28** (1976), 537–557.

[Ki] Kita, M., The Riemann-Hilbert problem and its application to analytic functions of several complex variables, Tokyo J. Math., **2** (1979), 1–27.

[Kl] Klein, F., Vorlesungen über die Entwicklung der Mathematik in 19 Jahrhundert, Chelsea P., New York, 1967.

[KS] Kimura, T. and Shima, K., A note on the monodromy of the hypergeometric differential equation, Japan. J. Math., **17** (1991), 137–163.

[Koh] Kohn Treibich, A., Un resultat de Plemelj, Progress in Mathem., **37**, Birkhäuser, 1983.

[Ko1] Kostov, V.P., Fuchsian systems on CP^1 and the Riemann-Hilbert Problem, preprint 303, the University of Nice, 1991.

[Ko2] Kostov, V.P., Fuchsian systems on CP^1 and the Riemann-Hilbert Problem, C.R.Acad.Sci. Paris, t. 315, Serie I (1992), 143–148.

[LD] Lappo-Danilevskii, I., Memoires sur la theorie des systemes des equations differentielles lineaires, Chelsea, New-York, 1953.

[Le] Levelt, A.H.M., Hypergeometric functions, Nederl. Akad. Wetensch. Proc., Ser. A **64** (1961).

[Lei] Leiterer, J., Banach coherence and Oka's principle, Contributions to the school on Global Analysis (Ludwigsfelde, 1976), P.2, Zentralinst. Math. Mech. Akad. Wiss. DDR, Berlin, (1977), 63–129.

[Lek] Leksin, V.P., Meromorphic Pfaffian systems on complex projective spaces, Math. USSR-Sb., **57** (1987), 211–227.

[Na] Nastold, H.J., Über meromorphe Schnitte komplex analytischer Vektorraumbundel und Anwendungen auf Riemannschen Klassen, Math. Z. **70** (1958), 55–92.

[Oht] Ohtsuki, M., On the number of apparent singularities of a linear dofferential equation, Tokyo J. of Math., **5** (1982), 23-29.

[OSS] Okonek, G., Schneider, H., and Spindler, H., Vector bundles on complex projective spaces, Birkhäuser, Boston, MA, 1980.

[Pl] Plemelj, J., Problems in the sense of Riemann and Klein, Interscience, New York, 1964.

[Poi] Poincaré, H., Sur les groupes des équations linéaires, Acta Math., **5** (1884), 201-312.

[Rö] Röhrl, H., Das Riemann – Hilbertsche Problem der Theorie der linearen Differentialgleichungen, Math. Ann., **133** (1957), 1–25.

[SJM] Sato, M., Jimbo, M. and Miwa, T., Holonomic quantum fields II: The Riemann-Hilbert problem, Publ. RIMS Univ. Kyoto, **15** (1979), 201–278.

[Sch] Schlesinger, L., Vorlesungen über lineare Differentialgleichungen, Teubner, Leipzig – Berlin, 1908.

[St] Steenrod, N., The topology of Fibre Bundles, Princeton, 1951.

[Su] Suzuki, O., The problem of Riemann and Hilbert and the relations of Fuchs in several complex variables, Lecture Notes in Math., **712** (1979), 325–364.

[Tr] Tricomi, F.G., Lezioni sulle equazioni a derivate parziali, Gheroni, Torino, 1954.

[Yo] Yoshida, M., Fuchsian differential equations, Vieweg Verlag, Wiesbaden, 1987.

Index

Harmonic Maps and Integrable Systems

Edited by Allan P. Fordy and John C. Wood

1994. viii, 330 pages. (Aspects of Mathematics, Volume E23; edited by Klas Diederich) Hardcover
ISBN 3-528-06554-0

From the contents: Introduction and background material – The geometry of surfaces – Sigma and chiral models – The algebraic approach – The twistor approach

Harmonic maps are maps between Riemannian or pseudo-Riemannian manifolds which extremise a natural energy integral. They have found many applications, for example, to the theory of minimal and constant mean curvature surfaces. In Physics they arise as the non-linear sigma and chiral models of particle physics. Recently, there has been an explosion of interest in applying the methods to integrable systems to find and study harmonic maps. This book brings together experts in the field to give a coherent account of this subject. The book starts with introductory articles by the editors, explaining what is needed to understand the other articles so that the book is self-contained. It will be useful to graduate students and researchers interested in applying integrable systems to variational problems, and could form the basis of a graduate course.

Vieweg Publishing · P.O.Box 58 29 · D-65048 Wiesbaden

vieweg

A History of Complex Dynamics

From Schröder to Fatou and Julia

by Daniel S. Alexander

1994. viii, 165 pages (Aspects of Mathematics, Volume E24; edited by Klas Diederich) Hardcover
ISBN 3-528-06520-6

From the contents: Schröder, Cayley and Newton's Method – The Next Wave: Korkine and Farkas – Gabriel Koenigs – Iteration in the 1890's: Grévy – Iteration in the 1890's: Leau – The Flower Theorem of Fatou and Julia – Fatou's 1906 Note – Montel's Theory of Normal Families – The Contest – Lattès and Ritt – Fatou and Julia.

The contemporary study of complex dynamics, which has flourished so much in recent years, is based largely upon work by G. Julia (1918) and P. Fatou (1919/20). The goal of this book is to analyze this work from an historical perspective and show in detail, how it grew out of a corpus regarding the iteration of complex analytic functions. This began with investigations by E. Schröder (1870/71) which he made, when he studied Newton's method. In the 1880's, Gabriel Koenigs fashioned this study into a rigorous body of work and, thereby, influenced a lot the subsequent development. But only, when Fatou and Julia applied set theory as well as Paul Montel's theory of normal families, it was possible to develop a global approach to the iteration of rational maps. This book shows, how this intriguing piece of modern mathematics became reality.

Vieweg Publishing · P.O.Box 58 29 · D-65048 Wiesbaden

Aspekte der Mathematik

Edited by Klas Diederich

*A Publication of the Max-Planck-Institut für Mathematik, Bonn